Genetic Engineering

Look for these and other books in the Lucent Overview series:

Genetic Engineering

by Clarice Swisher

LUCENT
BOOKS

To Emily and Laura

Library of Congress Cataloging-in-Publication Data

Swisher, Clarice, 1933–
    Genetic engineering / by Clarice Swisher.
        p.   cm. — (Lucent overview series)
    Includes bibliographical references and index.
    Summary: Surveys the uses of genetic engineering in medicine,
agriculture, and industry and the possible consequences of such
human interventions.
    ISBN 1-56006-179-0 (alk. paper)
    1. Genetic engineering—Juvenile literature.   [1. Genetic
engineering.]   I. Title.   II. Series.
QH442.S89   1996
575.1'0724—dc20   96–2676
                                                                CIP
                                                                 AC

Copyright © 1996 by Lucent Books, Inc.
P.O. Box 289011, San Diego, CA 92198-9011
Printed in the U.S.A.

Contents

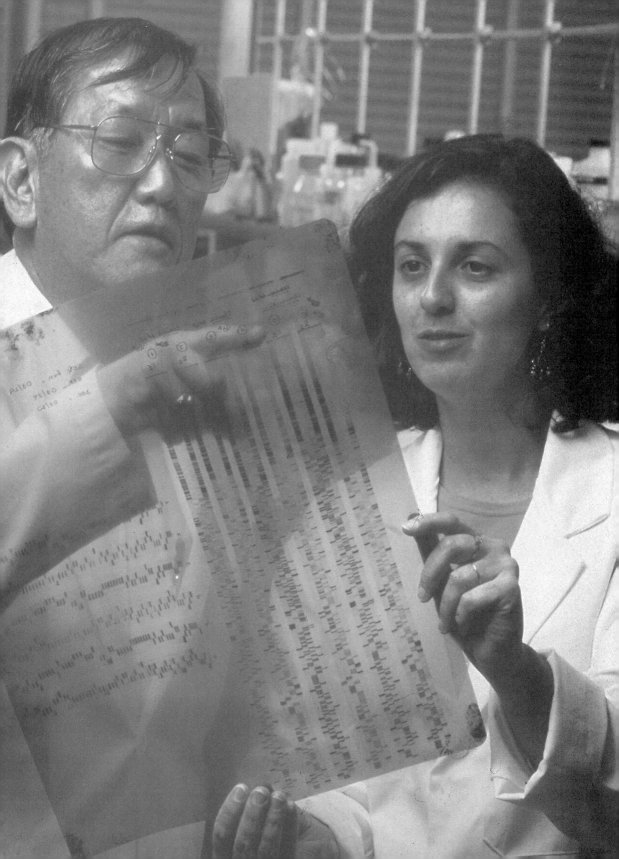

Introduction

GENETIC ENGINEERING is the deliberate alteration of an organism's genes by human intervention. The genes of every organism give that organism a particular set of traits and allow it to pass on the traits to a new generation. Since 1953, when scientists discovered the workings of the genetic system, scientists have learned how to change the composition of genes and how to change the basic traits of an organism. Having the technology to alter the way organisms grow and evolve has stimulated controversy and public debate on the ethics of genetic engineering and the effects of genetic engineering on the environment, agriculture, medicine, and industry.

Rumors about genetic engineering generate extremes of fear and hope. The fearful have visions of mad scientists creating bees the size of German shepherds and freak animals with two heads and five legs. They fear that scientists will secretly redesign the food supply to harm people in some unknown way. The hopeful believe that scientists can create an ideal world in which genetic engineering will create food to feed all the hungry, detergents to cleanse environmental toxins and wastes, and instant cures for terrible diseases. In reality, genetic engineering is poised to create neither the horrors of the fearful nor the miracles of the hopeful.

(Opposite page) Genetic engineers at work in an agricultural research lab. Genetic engineering promises to revolutionize agriculture, medicine, and industry. Many hope that scientists will soon be able to eliminate disease and create a healthier, more abundant food supply through genetic engineering. Some, however, fear the potentials of genetic engineering gone awry.

7

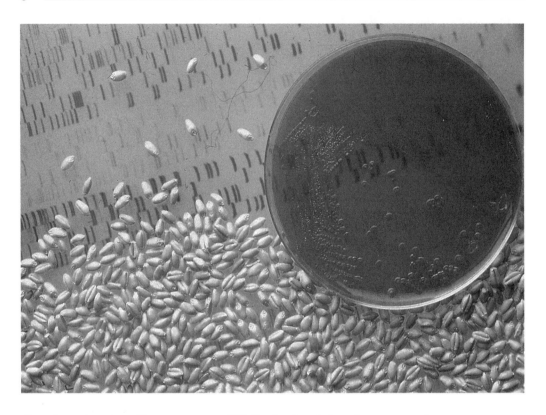

Genetic engineering offers myriad techniques to improve agriculture. Wheat seeds that are engineered with bacteria like the ones colonized in this petri dish resist disease.

While both of the above reactions are too extreme, many people do raise thoughtful questions about the basic concepts of genetic engineering. They see the possible danger of one group—scientists—holding the knowledge and power to redesign living things. Such power over the creatures of our planet seems too much like allowing a few people to play God. In addition, the public often wonders why taxpayers' money should go to research that produces seemingly useless products such as a polka-dot mouse or a fly with a dozen or more eyes. People wonder how much damage could occur if a harmful, engineered organism gets out of control.

Promises and concerns of genetic engineering

Genetic engineering indeed offers the possibility to improve agriculture—as long as scientists take adequate precautions. For example, genetic engineers know how to make plants more efficient and how to reduce crop losses. They know how to engineer animals to produce meat and milk more efficiently. These techniques enable farmers to produce more food for the world. On the other hand, altered bacteria that scientists use in the engineering process may produce undesirable results: They may mutate and regain their original harmful qualities and end up contaminating plant or animal food instead of improving it.

Much of genetic-engineering research since 1953 has focused on human health and medicine. Genetic engineering promises to diagnose, cure, and eventually prevent genetic diseases. These cures will become available after thorough testing proves that the treatments are safe.

Already scientists have developed many tests to screen, or examine, parts of human DNA for genes that are known to cause certain diseases. As more screening tests become available, there is concern that people will be coerced into having their genes screened. Such coercion would violate a person's right to privacy. In addition, if the screening discovers that a person has a disease-causing gene, there is concern that employers would refuse to hire them and that health insurance companies would either refuse to sell them insurance or charge them exorbitant prices.

Curing disease

Scientists have other concerns about using genetic engineering to cure disease. Scientists assume that one day doctors will have the knowledge and technical skill to identify a harmful mutated gene in a patient's cell and engineer it to be normal. If and when doctors have this technology, scientists want to prevent the engineering of germ-line cells because these cells contain the genes that control the traits a person passes to his or her descendants. Scientists believe that this new technology is too unrefined to risk the chance of adversely affecting future generations. On the other hand, an insurance company may want to engineer a mutated germ-line cell as a cost-saving measure. For example, if an embryo has a mutated cell that might cause cystic fibrosis, why not prevent it from appearing in the genome, or the complete genetic system, of the individual child and all future descendants of that embryo?

The newest field for genetic-engineering research is engineering microbes to perform environmental and industrial tasks. A primary example is the promise that engineered microbes can be used to clean toxic waste sites and oil spills.

On his hands and knees, a worker scrubs the oil-soaked rocks in Prince William Sound, Alaska, after the 1989 Exxon Valdez *oil spill. As a result of genetic engineering, microbes will soon facilitate the cleanup of such environmental mishaps.*

People ask many questions about corporate use of genetic engineering. A corporation may let its profit motive overrule its ethical responsibility and engineer a harmful product without concern for public health. In addition, scientists may be tempted to abandon basic research jobs in universities to engage in more profitable industrial research. Some people worry that industry will neglect essential, but unprofitable, research to solve human health problems and that government funds will be insufficient to keep these health projects alive.

Whether beneficial or harmful, genetic engineering will revolutionize the way we think about ourselves, the way we grow our food, and the way we care for ourselves and our environment. According to Noëlle Lenoir, president of UNESCO's International Bioethics Committee: "There are stages in the history of science when knowledge takes such great steps forward that our understanding of the world is completely transformed." The discovery of the structure of genes and the techniques to engineer them is one of these steps.

1

Genetic Engineering and Plants

SINCE THE BEGINNING of agriculture ten thousand years ago, farmers have been saving seeds and cross-breeding plants to produce the tastiest, hardiest, most abundant food crops possible. Cross-breeding occurs when the pollen from one species pollinates, or breeds, with a similar species. Records from ancient Sumeria also show that farmers planted vegetables among palm trees to protect the small plants from withering in the hot sun and saved the best plants for the next season's seeds. Food-growing techniques have progressed dramatically since those days, but the goal of producing the biggest yield from the smallest investment remains the same.

Today, genetic engineering techniques allow scientists to cross unlike species, such as two species of plants that do not naturally cross-pollinate. Using genetic engineering to combine genes from more than one species allows scientists to develop plants with new features. By inserting foreign genes, those from another species, into plant genes, scientists can improve the hardiness or nutritional value of existing species and develop new plants. Their goal is to reduce or eliminate the use of chemicals, to grow plants in conditions where they previously would not grow, to reduce spoilage and waste, and to increase production.

Today's engineered plants

Plants altered by a genetic-engineering process are already in the field and in the market. Wheat and rice plants have been engineered to increase yield by producing shorter, stiffer stalks that help the plants survive and grow in severe weather. Gene-altered rice occupies several million

A researcher examines rice plants that have been genetically engineered to be more nutritious.

hectares (a measure of land comparable to nearly two and a half acres) in China, a large increase in that country's most important food. Besides producing hardier crops, scientists have engineered plants that resist insects. Geneticist Abhays Dndekar of the University of California at Davis has engineered walnut and apple trees that resist the coddling-moth larva, which develops into a worm that lodges in the apple and the walnut. Other scientists have modified cantaloupe, melons, and cotton plants to resist insects. At the University of Wisconsin at Madison, geneticist Brent McCowan has introduced a gene from a bacterium into poplar trees, making the trees toxic to caterpillars that eat the young saplings. At the University of California at San Diego, geneticist Maarten J. Chrispeels has successfully inserted a gene from kidney beans into peas. Weevil larvae that ordinarily nibble on peas in storage bins dislike the new variety. Chrispeels thinks it possible to use the same technique to decrease weevil loss in cowpeas, black-eyed peas, and mung beans, all of which are important protein sources for people living in Africa, Asia, and South America.

Disease-resistant plants

Other plants have been engineered to resist diseases, the process that scientists believe will ultimately achieve the biggest savings in food production. Virus-resistant squash were marketed in 1995, while resistant cucumbers and cantaloupe are scheduled for production in 1996 or 1997. Disease-resistant corn, soybeans, cotton, tomatoes, and potatoes have already been produced and are being tested prior to marketing. At the Scripps Research Institute in La Jolla, California, a team of plant biologists led by Roger N. Beachy has engineered rice with extra genes so

Plant biologist Roger N. Beachy has engineered virus-resistant rice. Healthier crops will mean increased food production.

that it resists a destructive virus disease. This rice is currently being field-tested, and if the tests are successful, the rice will be marketed around the year 2000. Scientists have developed other genetic-engineering techniques to reduce spoilage, a factor especially important today when fresh foods are shipped around the world. The first genetically engineered food, the FLavr Savr tomato, has already appeared in the supermarkets. The FLavr Savr contains a new gene that slows the production of an enzyme that causes the tomato to soften. By slowing this enzyme, the tomato remains firm longer and does not spoil as quickly. In commercial terms, the tomato has a long shelf life.

Engineered plants for the future

Scientists are conducting a variety of research projects to improve future plants. The goal of some projects is to reduce insect damage. To give plants their own built-in pesticide, scientists are trying to engineer a toxin that is lethal to insects and add it to crop plants at the gene stage. Scientists have already found that by inserting a wheat gene into corn, they can make the corn taste noxious to insects that normally thrive on corn.

The goal of other projects is to reduce disease damage. Developing plants that resist diseases is especially important in tropical areas with hot, wet climates that invite rapid spoilage. According to the U.S. Department of Agriculture (USDA), scientists are working on engineering a gene from a chicken to make potatoes more resistant to diseases. Other scientists are trying to insert a protein from the greater wax moth into a potato to make the potato less likely to bruise. Other projects in progress may keep bananas and strawberries firm

Genetically engineered FLavr Savr tomatoes are designed to stay fresh longer than ordinary tomatoes. They contain a gene that slows the production of an enzyme that causes spoilage.

longer, produce cooking oils with less fat, and create nuts that do not turn rancid or cause allergic reactions. One researcher is trying to engineer a coffee bean without caffeine in order to eliminate the need to decaffeinate coffee.

Scientists know that certain bacteria can convert nitrogen from the air where it is plentiful; farmers add nitrogen to soil as fertilizer. One group of scientists is trying to engineer nitrogen-converting bacteria into wheat plants so that the wheat can produce its own fertilizer—just as soybeans, clover, and alfalfa do naturally. If scientists can engineer this nitrogen-converting bacteria into the wheat plants, farmers will be able to grow wheat without using commercial chemical fertilizers, which run off into the water supply. In another project, researchers are trying to engineer blue-green algae into the structure of fern plants. Blue-green algae can also convert nitrogen from the air. The plan is to use ferns as fertilizer factories in rice fields.

Growing plants in unlikely places

Other projects still in the research stage are engineering plants to grow in new conditions. For example, scientists have developed a genetically engineered bacterium that protects plants in temperatures as low as twenty-three degrees Fahrenheit. Scientists have been experimenting with growing frost-resistant broad leaf plants, such as strawberries. The ability to grow freeze-resistant strawberries would increase production because farmers would have a longer growing season and strawberries could grow in colder climates. This bacterium must be tested, however, for different soils, soil treatments, and weather conditions before it is approved for widespread use. Other projects are under way to engineer plants to grow in places where they do not ordinarily grow. Be-

cause only a few plants thrive in conditions where the soil has high acid or high salt content, researchers are working to develop a gene pool from plants that naturally grow in these conditions so that they can insert the hardy genes into crop plants. In one project, scientists are experimenting with mangrove trees that grow in tidal regions and can tolerate saltwater conditions. Scientists hope that engineering crop plants with mangrove-tree genes will allow crop plants to grow in soils with high salt content.

Mich Hein, who works at the Scripps Research Institute in La Jolla, California, and other researchers around the world are working to engineer genes into plant foods so that the plants can be used as vaccines and drugs. On May 5, 1995, Warren E. Leary reported in the *New York Times*, "The research raises the possibility that vaccines

Genetic engineering has become an important tool in the production of vaccines. Cell biologist Mich Hein is working on altering genes in alfalfa so that the plant can be used as a vaccine against the intestinal disease cholera.

delivered by fruits and vegetables might eventually replace those given by needle." Once such plants have been engineered, pharmacological scientists could extract the vaccine and make it into pills or people could eat the vaccine plants. Hein is altering alfalfa to carry a gene to inoculate humans against cholera, an intestinal disease. Molecular biologist Andrew Hiatt is trying to engineer alfalfa so that it manufactures antibodies that prevent tooth decay. Charles Arntzen of Texas A&M University is engineering bananas to be used as vaccines against hepatitis B and intestinal diseases. At Virginia Polytechnic Institute, researcher Carol Cramer is developing blood-thinning drugs by engineering tobacco plants.

Non-polluting and inexpensive

Researchers believe that plant vaccines and drugs have several advantages over other drug-delivery methods. One advantage is that plants get their energy from the sun; therefore, their production does not pollute the environment. In addition, growers can produce plants inexpensively and eliminate the cost of manufacturing a pill. Another advantage of plant vaccines and drugs is that plants can be produced where they are needed. Bananas, which can be eaten raw, are a particularly good fruit to use for vaccines and drugs because they grow abundantly in underdeveloped countries that desperately need cheap, available vaccines. Experimental plots of genetically altered bananas should be growing before the end of the century. Major issues remain, however, before plants replace pills in the drugstore: Researchers must conduct extensive tests to see how the immune system responds to the drugs, how big a dose a single plant or fruit can produce, and how long the plant retains the engineered gene.

Genetic engineering also enables scientists to duplicate, or clone, plants. Cloning is a process of reproducing identical copies by using a bacterium that reproduces itself rapidly. As the bacterium reproduces, so do the recombined genes from plants. This process allows growers to produce new varieties of plants. Scientists can select flower genes for blossom size and color and modify the colors and sizes as they choose. Because cloned plants grow faster and are more uniform than plants grown from seed, nurseries prefer to buy many varieties of cloned plants, from ferns to strawberries.

Genetic engineering of plants holds out the possibility of increasing the world's food supply and reducing the cost of production. Plants that withstand weather and resist insects and spoilage reduce the loss of much-needed food. Plants that have the capacity to withstand larger applications

The goal of some genetic engineering research is to make plants less susceptible to harmful insects. Unlike the untreated leaf on the left, the genetically engineered cotton leaf on the right has resisted insect damage.

of herbicides and that resist shipping damage reduce loss and increase the supply. On the other hand, many researchers express valid concerns about the possible harmful side effects of these techniques.

Concerns about harmful effects

Some molecular biologists, the scientists who conduct genetic research, worry that various species of bacteria and viruses used in genetic engineering may cause harm. For example, a lethal bacterium inserted into corn genes to allow the plant to manufacture a pesticide that kills a particular insect may also kill other helpful insects. Liebe F. Cavalieri, environmental science research professor at the State University of New York, said in an article in the December 1994 issue of the *World & I*, "Anything that kills one form of life is likely to have an effect of some kind on other forms." To prevent serious consequences to other life forms, some scientists and many lay people want to continue experimentation and delay distribution of these altered plants until the long-term consequences of genetic engineering are known. The consequences, however, cannot be known for certain without field testing. Robert Bohrer articulates the researchers' frustration in "The Future Regulation of Biotechnology": Scientists are restricted by "a regulatory framework in which nothing could be done [grown] because nothing can be proven safe [tested] and nothing can be proven safe because nothing can be done." Scientists have, however, received government permission for controlled field tests for a number of projects.

Dangerous mutations

Harmful mutations may be another side effect of genetic engineering. Plants genetically engi-

neered to resist herbicides and insects may mutate into strong strains in two ways. First, plants that resist herbicides and insects are less likely to be destroyed and, therefore, could mutate and become hardy enough to choke out surrounding plants. Second, herbicide- and insect-resistant plants could cross-pollinate with weeds of a related species. In this process, the weeds could acquire the plants' herbicide- and insect-resistant traits and become even stronger and harder to control.

Other harmful side effects may result from a bacterium gone awry. Scientists are concerned that the bacteria and viruses used in the engineering process may have harmful and unexpected consequences. In one stage of the engineering process, scientists must use a bacteria or virus as a carrier because it can replicate itself without a mate. A carrier is a bacterium that "carries" the genes from one species to another. After the genes from two species are dismantled, they recombine and the bacterium that replicates quickly without a mate makes many copies of the recombined genes. Although researchers remove harmful genes from these bacteria and viruses before using them, the possibility exists for harmful genes to return in the future, spread into the environment, and threaten the health of humans and animals.

Other concerns involve unintended but potential risks. For example, plants naturally produce harmless low levels of toxicants, or potentially poisonous substances. Genetically engineered plants may produce natural toxicants at higher, harmful levels. Another unintended effect of genetic engineering may be a change in the levels or structure of a plant's nutrients; that is, a plant's nutritional value may change—oranges, for example, without vitamin C. Moreover, a plant may acquire allergens (allergy-causing substances) in the engineering process. If a gene from an allergy-causing plant is inserted into a plant that does not have the same allergens, it may acquire them. An allergy-sensitive person could experience a dangerous reaction to foods that were formerly safe. These concerns increase the need for thorough testing before the altered plants are approved for public consumption.

The ecologists' concerns

The concerns of ecologists about genetically engineered plants are even more extensive than the concerns expressed by molecular biologists. Ecologists, who consider the entire ecosystem, foresee several potential dangers. If given a head start, an engineered organism may spread, drive out natural plants in an area, and consequently, affect other species that depend on the natural plants for food. Moreover, ecologists remind advocates of genetic engineering that no one can predict or control mutations. An engineered organism may cause harmful mutations in existing organisms, and scientists cannot ensure that such mutations will not happen. Ecologists also remind advocates that evolution is irreversible; that is, no species ever retraces its evolutionary steps. In other words, if a harmful mutation occurs,

Molecular geneticist Peter Quail examines greenhouse-grown rice plants. Most genetic researchers agree that genetically engineered plants will not threaten the health of humans and animals.

there is no way to undo the harm. That fact causes scientists David Suzuki and Peter Knudtson to be cautious. In their book *Genetics: The Clash Between the New Genetics and Human Values*, Suzuki and Knudtson caution that no one is able "to reliably anticipate the evolutionary consequences of releasing novel genetically engineered life forms into nature."

Assurances from genetic researchers

Most genetic researchers, however, say that genetically engineered plants are no more dangerous than plants whose mutations occur in nature. Furthermore, researchers say that genetically engineered genes, which are artificially made mutations, are weaker and more likely to disappear within a few generations than natural mutations. In *The Genetic Revolution: Scientific Prospects and Public Perceptions*, author and editor Bernard D. Davis quotes the U.S. National Academy of Sciences' policy statement concerning the introduction of engineered organisms into the environment:

> There is no evidence of the existence of unique hazards either in the use of recombinant DNA techniques or in the movement of genes between unrelated organisms.

> The risks associated with the introduction of recombinant DNA-engineered organisms are the same in kind as those associated with the introduction of unmodified organisms and organisms modified by other methods.

Davis agrees with the academy's position and sees no logical reason for treating genetically engineered organisms differently from cross-pollinated hybrids. Nevertheless, he says, "the powerful social tendency to impose stricter scrutiny and control over new technologies may prevail over logic."

"FIRST THEY DECLARE THE EARTH IS ROUND, AND NOW *THIS!*"

All scientists, the biologists who advocate genetic engineering of plants and the ecologists who oppose it, agree on one issue: that reducing chemicals in the environment is a desirable and necessary goal. But they disagree on how genetic engineering will affect that goal. Advocates argue that genetically engineered organisms are likely to cause minimal harm to human health and will greatly reduce the need for chemicals. Prolonged testing delays the use of safe, genetically engineered plants, they believe. The delay means the use of more chemical fertilizers and pesticides that harm the health of the public and the farmworkers and pollute the environment. Therefore, advocates of genetic engineering argue that delays increase environmental pollution. Ecologists, however, fear that some genetically engineered plants will cause more chemical pollution.

This genetically engineered barley carries a gene that helps the plants resist disease. Researchers hope that plants like these will help reduce the amount of chemicals used in agriculture.

Herbicide-resistant plants can withstand larger doses of weed killer and will encourage chemical companies to use even more chemical herbicides. Ecologists note that the same companies sell both the herbicide-resistant seeds and the herbicides to use with them. Ecologists argue that these companies can make a double profit, and the environment is no cleaner. In the end, the debate becomes a polarized "struggle between those interested only in profit and those who would defend the environment against all profiteers," according to ecologist Simon A. Lavin, who wrote "An Ecological Perspective," published in *The Genetic Revolution: Scientific Prospects and Public Perceptions.*

Despite the differences of opinion among scientists, the gene revolution in agriculture is likely to go forward at a rapid pace. Many scientists predict an explosion in plant engineering in the near future. Most biologists and geneticists who conduct plant research are excited and hopeful. They visualize hardier, more efficient plants growing on formerly unproductive land with fewer chemicals. They see greater production of food. They see valuable land and the environment protected for wildlife and for people. However, the future is still a great unknown. In 1992 Vice President Al Gore suggested that since genetic-engineering technology is developing so rapidly, there is too little time to explore, study, and debate enough. Americans, he said, may have to make some tentative decisions "without the base of understanding that a democracy requires for difficult decisions."

2

Genetic Engineering and Animals

ANIMALS PLAY A MORE extensive role in genetic engineering than plants play. Scientists engineer plants to increase production, reduce loss, and reduce the use of chemicals as part of the broad goal of improving the food supply and the environment. Scientists also engineer animals in several ways to improve the food supply. In addition, they use animals for basic research. Since the discovery of DNA, the chemical system making up the genes of all living organisms, scientists have used animals to explore the intricate workings of the genetic system. They in turn use that knowledge as the basis for developing genetic engineering techniques. In 1991, for example, researchers used between eighteen million and twenty-two million animals. Fifteen million to nineteen million of them were mice and rats bred specifically for research purposes.

The use of genetically engineered animals offers great research possibilities and causes widespread public concern. The many possibilities for increasing food production and reducing losses in animal agriculture are balanced by the potential

(Opposite page) Researchers, who use millions of animals each year in basic research, are also developing techniques to genetically engineer animals to increase livestock productivity and increase the food supply.

31

problems. Because scientists have no absolute way to guarantee that bacteria used in the engineering process is totally stable when an animal's genes are altered, health safety and integrity of the food supply are the greatest concerns. Other concerns relate to basic research. People worry that scientists are probing too much into the secrets of life and wonder about the practical value of some experiments. In addition, people question the ethics of using animals for experimentation.

Animal research in agriculture

Genetic engineering promises to increase livestock productivity by generating more meat and milk from a smaller investment in feed and animals. To reduce the loss of farm animals, scientists have genetically engineered vaccines that prevent diseases. In the past, infectious diseases in herds were a problem that farmers solved by slaughtering both sick animals and healthy ones, which were sure to become infected. Later farmers added antibiotics to animal feed. Today, vaccines genetically engineered with a resistant bacteria keep newborn pigs and calves from contracting diarrhea and hoof and mouth disease. Vaccines eliminate both the need to fortify feed with costly antibiotics and the risk that humans could ingest excessive amounts of antibiotics through milk and meat and become immune to antibiotics. Vaccines, however, increase the risk that resistant bacteria meant for animals could spread to humans through meat or milk products.

Growth hormones

Feeding cattle growth hormones is another practical application of genetic-engineering technology. Scientists can produce a large supply of any organism's natural growth hormones by cloning them. Injecting cloned hormones into cat-

tle increases the quantity of milk and meat. Late in 1993 the Food and Drug Administration (FDA) approved the use of a bovine growth hormone, bovine somatotropin (BST). BST speeds the growth of beef cattle and increases milk production of dairy cattle by 10 to 15 percent or more. The hormone reduces production costs because cattle produce more meat and milk without additional feed.

The public has been wary of accepting milk from cows given growth hormones. When the milk from BST-fed cows first appeared in the supermarkets, the Pure Food Campaign protested by dumping milk outside stores in more than a hundred American cities. The protesters insisted on the right to know if a specific quart of milk had been produced by genetically engineered hormones. Proponents of the hormone say there is no evidence that the milk is harmful and that milk from cows given BST is almost identical to milk

The appearance in supermarkets of milk from BST-fed cows was met with public resistance in some areas. Here, activists protest against a company that uses BST by dumping milk on the streets.

produced naturally. Moreover, proponents claim that 90 percent of BST is inactivated when milk is pasteurized and meat is cooked. Opponents note, however, that cows given BST develop oversized milk bags and udders and suffer from a high incidence of udder infections, which are then treated with antibiotics that might get into the milk.

Farmers also use growth hormones to increase productivity and quality in pigs. Porcine, the growth hormone for pigs, increases weight gain by 10 to 20 percent. When pigs grow faster, they produce more meat with less fat and use less feed. Feeding animals with genetically engineered growth hormones is an international practice. An estimated ten thousand pigs in the United Kingdom receive growth hormones annually. There is a limit, however, to the amount of growth hormones that can be given to pigs and cattle. High dosages cause animals to grow so large that their legs are too weak to hold up their enlarged bodies, and they develop injuries. In an article in the *Animals' Agenda*, Andrew Kimbrell, executive director of the International Center for Technology Assessment in Washington, D.C., cited the example of pig No. 6707. Pig No. 6707 was supposed to turn into a superpig from injections of growth hormones. Instead:

> The growth hormones altered his metabolism in an unpredictable and unfortunate way, and he turned into a tragicomic creation: an excessively hairy, lethargic [lacking energy to move], apparently impotent, slightly cross-eyed pig, riddled with arthritis, who rarely stood up.

A pig who received growth hormones (left) shares a pen with a smaller pig whose feed was not supplemented with hormones.

Besides using hormones to stimulate cattle and pig growth, farmers also feed growth hormones to chickens. Growth hormone reduces the time it takes a chick to grow to broiler-size by 15 percent. A new hormone for egg-laying chickens prevents hens from molting, or entering a natural rest time during which they do not lay eggs. Australian ranchers use hormones to produce sheep that grow 30 percent faster. Using hormones to stimulate livestock growth cuts the cost of food production all over the world.

Animal research offers future possibilities

Although many genetic-engineering techniques have already been put into practice, others are still in the research stage. Scientists are conducting research on the use of foreign genes to produce new features in animals. For example, researchers in Australia are working on transplanting genes to make sheep's wool grow faster and to make sheep drop their wool at a particular time, thus eliminating the need for shearing. Researchers in Canada are inserting human, chicken, and cattle growth hormones into salmon to force them to grow larger. In the future, pigs and cattle may produce less saturated fat, and chickens may lay eggs with less cholesterol.

Researchers are also experimenting with cloning. The traditional methods of crossbreeding can produce prime animals, but improving a breed takes generations of natural growth because cows produce one calf at a time. By cloning, researchers think they can produce several identical copies of a superior animal from one embryo. To clone a calf, researchers remove an early-stage embryo from a pregnant cow, separate the cells, remove the nucleus from each cell, insert each nucleus into a new egg, and implant the altered eggs into surrogate (substitute) females. From one embryo, scientists can obtain from four to eight identical copies of the animal. This cloning technique has several advantages over natural crossbreeding. Cloning one superior embryo can produce several calves a season instead of just one; dairy farmers can select the superior females and beef farmers the superior males. Eventually, scientists expect to engineer the embryo with foreign genes so that they can add desirable traits to the DNA before they clone it.

Fish are easier to clone than cows. Because fish develop from eggs outside of the mother's body, and, therefore, do not need surrogate mothers, scientists can easily inject new genes into the eggs before cloning them. With the rapid depletion of the fish supply in northern Atlantic waters, some scientists predict that a portion of the fish demand in the future will be filled by fish cloned on farms.

A common concern runs through all engineering of animals for agricultural purposes. This concern is that bacteria used in the process may become unsafe and cause harm to people's health.

Using animals in basic research

Many people are confused about the use of animals in basic research. The media report spectacular results, but they often fail to give a thorough explanation of the purpose of the research or tell how the experiment added new knowledge to the fields of molecular biology and genetic engineering. Because the genetic system is very complex and engineering genes is a new and intricate process, scientists, especially in university laboratories, have to learn one small step at a time. They learn about genes and how to engineer them by conducting basic research, most often using animals.

Many basic experiments have produced transgenic animals, or animals that are modified with foreign genes. These experiments help scientists learn the procedures for injecting genes from one animal into another. Several researchers focused on genes that control an animal's size. For example, rat genes injected into one white mouse in a litter produced a mouse the size of a rat while its littermates remained normal size. Another researcher used rabbit genes to produce a bigger mouse. A more complicated project produced a white mouse with black dots. Researchers injected human DNA into cells from the embryo of a white mouse and cloned them. From the resulting gene library, the millions of recombined cells that result from cloning, they selected the white mouse cell that contained human DNA. They injected this mutated cell into the embryo of a black mouse and grew the embryo in a surrogate mother. The result was a white mouse with black dots carrying human DNA. Scientists had performed two important steps: They had added foreign genes, and they had combined the genes of two mice. Transgenic animals in basic research are not produced for practical use; obviously, there is not much demand for extra-large or polka-dot mice. The purpose of the laboratory models is to contribute to the knowledge of genetics and genetic engineering.

Some of the projects are designed to develop new techniques. Researchers conduct thousands of experiments in which they refine techniques to add, delete, recombine, edit, and insert genetic material. One researcher produced a hairless mouse by deleting its hair genes. At the University of California at Davis, scientists studied a new transgenic process, a technique called cell fusion, in which they fused cells from sheep and goats, two unlike species. The outcome of their

This hairless mouse was the result of an experiment in which a researcher deleted the mouse's hair genes.

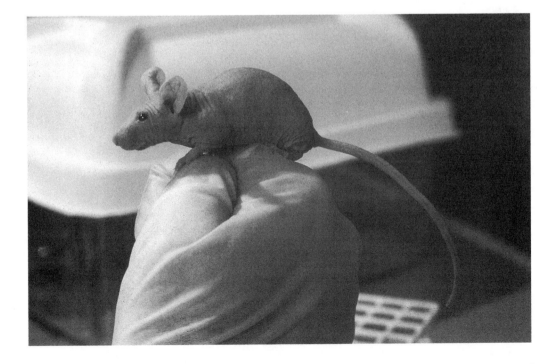

project was a creature that had the face and horns of a goat and the body of a sheep. Again, creatures like hairless mice and sheep-goat combinations exist only to advance knowledge and refine techniques.

A Tokyo research team experimented to see if they could transplant genes into a fetus. Scientists at the National Cancer Center Research Institute and the Chiba University School of Medicine, both in Japan, recently reported the results of an experiment in which they tried to alter the genes in an embryo while it was in the womb of the mother. They wanted to alter the genes of one fetus, but not affect future generations. They injected altered genes into the tail veins of pregnant mice and found that the newborn mice also carried the altered genes. Because the altered genes were no longer present after fourteen months, researchers knew they had not altered the genes that pass on traits to future generations. Though the technique may someday have practical value, Dr. Masaaki Terada, a member of the research team, said the purpose was simply to discover if the technique would work.

"Eyeless"

A team of Swiss scientists from the University of Basel, Switzerland, wanted to know if there were genes to control large body parts, such as limbs, liver, and brain. In March 1995 they reported on their search for genetic signals that initiate growth of the body's large components. Using fruit flies, the team reported that they had engineered flies with as many as fourteen perfectly formed eyes that grew on the flies' wings, legs, and the tips of their antennae. This team of scientists may have discovered what they call "the master control gene" for the formation of the eye. The scientists call the gene "eyeless" be-

A scanning electron microscope reveals an extra eye (indicated by the arrow) on the head of the genetically engineered fruit fly with the gene "eyeless."

cause the absence of it results in flies with no eyes at all. The senior author of the report, Dr. Walter J. Gehring, said, "It's an amazing example of how a single gene can switch on an entire developmental program. . . . It came as a total surprise to us." Dr. Gehring estimates that at least twenty-five hundred different genes participate in the construction of the eye and that all of those genes take directions from eyeless.

The scientists see no practical use for their newly created fly, but eventually scientists may be able to use the discovery to devise therapies for solving human visual problems. Since the report on the isolation of eyeless, scientists in Edinburgh, Scotland, have set up a project to try to fuse the fly eyeless gene into mouse genes. They hope they can correct a genetic disorder that causes eye defects in mice. In addition, the Swiss team has already planned experiments to see if the extra fly eyes are connected to the brain and if

Rats in a genetic-engineering research lab. Opponents of genetic engineering believe that it is morally wrong for scientists to alter the genes of any animal. Those who favor genetic engineering, however, think the knowledge gained through research outweighs any disadvantages.

they can see. The discovery made from the eyeless experiment marks a forward step in knowledge about the genetic system: Scientists have evidence of a single gene that controls a whole unit, like the eye. This knowledge may help scientists as they try to discover if genes govern other units or govern behavior.

Concerns about genetic engineering of animals

The use of animals for basic research raises moral and ethical concerns based on people's values and religious beliefs. Those who believe that God is all knowing and holds the secrets of life think scientists are going too far when they explain life according to genes. They think it is wrong for scientists to use their ability to alter genes to control the destiny of living creatures. One of the most outspoken opponents of genetic engineering, Jeremy Rifkin of the Foundation on Economic Trends, said:

> We are embarking on a very potentially troublesome journey, where we begin to reduce all other animals on this planet to genetically engineered products. . . . We will increasingly think of ourselves as just gene codes and blueprints and programs that can be tinkered with.

Since genetic engineering requires a great amount of scientific expertise, only a few scientists have the ability to alter genes. Opponents of genetic engineering fear that even an ethical scientist could have an accident and that an unethical one might deliberately misuse his or her skill.

Those who favor the use of animals in basic research believe that the knowledge gained outweighs the disadvantages. Moreover, they see no unusual moral or ethical dilemma. They point out that humans have always been curious and have always worked hard to discover the secrets of life and the universe. Discoveries involving genes are no more unusual than past discoveries about how the heart pumps blood through the body. In *Introduction to Genetic Engineering*, author William H. Safer said that genetic engineering is "an exciting, intellectually stimulating enterprise; an area of study that is bound to accelerate our understanding of how living things work."

3

Genetic Engineering and Medicine

IN THE IDEAL WORLD of medicine, scientists would be able to identify disease-causing genes, correct them, and prevent illness. Even though that goal seems impossible today, it may be achievable in the distant future. In the interim, scientists are using genetic engineering to make progress in four aspects of medicine: identification of genes that cause diseases; screening for diseases in individuals; treatment of diseases with engineered drugs; and gene therapy, or the genetic engineering of sick people's genes. These four areas of research are progressing at a phenomenal rate, as medical geneticist Dr. Albert B. Deisseroth of the University of Texas M. D. Anderson Cancer Center in Houston said: "[As late as 1992] I don't think any of us in the field thought it would move so fast." Nobel laureate David Baltimore, a molecular biologist at New York's Rockefeller University, said in 1993, "Every disease we know about is either being attacked with genetics or is being illuminated through genetics."

In 1958 scientists had identified approximately four hundred genetic disorders; by 1994 they had

(Opposite page) Deciphering the human genetic code could help researchers identify disease-causing genes, alter them, and prevent illness.

identified between four and five thousand. Most genetic disorders can be traced to a single defective, or mutated, gene. Some genetic disorders, however, involve multiple genes, and some diseases have a genetic component but are not totally controlled by one gene or a collection of genes. To find a defective gene, scientists use a variety of genetic-engineering techniques. One technique, called a probe, is a chemical that the desired gene will form a bond to; that is, one chemical sequence will attract another, in some ways as a magnet attracts metals. Scientists also identify markers, or known genes, among the chromosomes, the strands of DNA containing a number of genes, and search for a defective gene nearby. Another technique for locating defective genes uses antibodies. To study a defective gene and search for a cure, the researcher must first locate the mutated gene causing the genetic disorder.

The human genome

Identifying defective genes in the human genome—the complete human genetic system— means searching for one biochemical protein among millions of biochemical proteins, called nucleotides. The human genome has forty-six chromosomes, which reside within the nucleus of every body cell except blood cells. Of the forty-six chromosomes, forty-four occur as matched pairs in which one chromosome is inherited from each parent, as are two additional gender chromosomes. Though the full complement of human chromosomes is always forty-six, the amount of DNA in each chromosome varies. The long, double strand of DNA, called the double helix, contains an estimated 100,000 genes of varying sizes. Genes are nature's way of partitioning the DNA double helix into units, each of which has a specific role to play in the cell's activity. Genes are

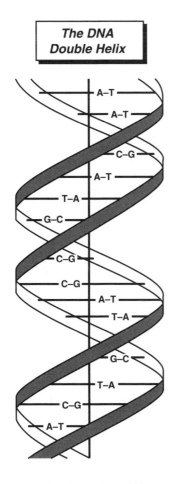

The DNA Double Helix

composed of nucleotides in different sequences. Nucleotides are composed of four bases, represented by the letters A, G, C, and T, that make up the genetic code. Identifying a gene means discovering the sequence of base pairs. They are called pairs because one base comes from each of the two strands of the DNA double helix. Some genes have fewer than ten thousand base pairs, and others have more than two million. Genes function both alone and in families. Before engineering a gene, a scientist has to locate the right one on the double helix. Finding a gene in this invisible mass of biochemicals may seem impossible, but scientists have developed efficient search techniques. In 1991 a machine that can sequence thousands of nucleotides daily was developed, and new machines being developed promise to work even faster.

Tracking down defective genes

Scientists have identified the defective genes for a number of serious diseases and new defective, or mutated, genes are detected every year. Scientists first located the gene for sickle-cell anemia, a disease that occurs when red blood cells form hard, sharp crescents, or sickles, that cause severe pain as the blood circulates. Scientists have also identified the gene responsible for one of the most prevalent genetic diseases, cystic fibrosis, in which mucus fills the lungs and interferes with breathing. They have found the gene for Huntington's disease, a disorder that causes progressive deterioration of the nervous system in adults, and the gene for Down's syndrome, a congenital condition including mental retardation. Scientists recently discovered that several mutated genes, rather than a single gene, cause Lou Gehrig's disease, a fatal disease that destroys the nerves. After years of research, scientists have identified the gene for Alzheimer's disease, which causes loss of memory in older people. They have identified the gene that causes Parkinson's disease, which destroys brain cells and causes muscle tremors, facial paralysis, and difficulty with walking and sitting. One of the most recent discoveries is a gene that produces an excess of protein in diabetes patients.

Scientists continue to search for genes that cause cancer, which is a disease that has a genetic component, but is not identified strictly as a genetic disease. Cancer results from an extreme breakdown in the normal processes that regulate growth; cell division goes on with fewer and fewer checks and balances until the body is destroyed by malignant cells. The genes that provoke cancer (some forty different ones discovered by 1987) are called oncogenes, and each one carries a different virus. In research labs, scientists can make oncogenes by genetically engineering them, and they can transplant oncogenes into animal cells for research. Thus far, scientists have identified the gene for malignant melanoma, a cancer that causes tumors that originate in the skin; the gene that causes leukemia, the most common kind of childhood cancer; and two genes that cause a small portion of breast cancers. Each new discovery, no matter how small, contributes to a broader understanding of the disease.

Screening and diagnosis

Once scientists have identified a gene that causes a physical genetic disease, they can screen, or test, for its presence or absence in the genome of an individual person and make a diagnosis. Traditionally, doctors have diagnosed diseases by observing patients' body signs and by asking patients about their symptoms. DNA technology is revolutionizing traditional diagnosis now that laboratory technicians can detect minute deviations in a patient's genes and chromosomes. In *Reshaping Life: Key Issues in Genetic Engineering*, G. J. V. Nossal said, "We can examine a person's individual genes with great scientific accuracy."

Using genetic engineering, scientists have developed screening devices, or tests, for diagnosis. From a small sample of body tissue or even the root of a single hair, doctors can analyze a patient's DNA. A complete analysis of a person's DNA is called a genetic fingerprint. From it, scientists can discover mutated genes that cause genetic diseases and immune disorders. Screening techniques can also detect viruses that other tests miss because the viruses are hidden inside the cells. An example of a hidden virus is the chickenpox virus, which can be dormant for years before it reemerges as shingles, an inflammation of nerves.

Doctors can screen an individual genome even though no two individuals, except identical twins, have the exact same gene pattern. English scientist Dr. Alec Jeffreys developed the system that detects the unique patterns. He found that the human genome has short sequences of nucleotides that are repeated hundreds or thousands of times at different positions in the DNA. The variations in position and length of these repeated sequences make each person's genome unique. Be-

Scientists separate fragments of DNA from blood cells into bands that will form a genetic map that is unique to each human being or animal.

Mark R. Hughes was part of a team that developed a genetic test to detect the presence of cystic fibrosis before birth.

cause the genome of all people is similar in most ways, screening devices need not be designed for each individual.

At first, screening for disease had to be done by scientists who alone could identify the gene being screened for. As the genes for more and more diseases are identified, corporations have begun to produce commercial tests to screen for genetic disorders. Already a test for cystic fibrosis is on the market. In March 1995 OncorMed Incorporated, a biotechnology company in Gaithersburg, Maryland, advertised that it could test people for the presence of genes that might increase their risk of developing breast cancer, melanoma, thyroid cancer, and brain tumors. People who test positive for such genes can then pursue more frequent screening and a better chance for early treatment.

Identification of the genes for all the possible disorders and a thorough map of the human genome will give scientists the knowledge they need to make accurate diagnoses. "Genetic engineering . . . will alter the landscape of diagnostic procedures so that, in twenty years' time, it will be virtually unrecognizable," predict medical researchers G. J. V. Nossal and Ross L. Coppel in the second edition of *Reshaping Life: Key Issues in Genetic Engineering.*

Researchers are also searching for genes that control human behaviors, such as genes that increase a person's risk for alcoholism, mental illnesses, and being overweight. A conference originally scheduled to take place in 1994 to discuss research to find a gene that increases a person's risk to act with violence had to be delayed for a year, however, because objections to the studies were intense and well publicized. Many people feared that discussion of a gene increasing one's tendency toward violence would make a connection between violence and race.

The distinction between a gene that causes an outcome and a gene that increases the risk for an outcome is very important. A person who has a gene that causes something will get whatever it causes. A person who has a gene that increases the risk for something may never get whatever the gene might have caused. So far, researchers think that genes may increase a person's chances for a designated behavior, but genes have not been found to cause a behavior.

Genetically engineered treatments

Genetic engineering has also produced treatments for diseases not caused by mutated genes. Researchers have used genetic engineering to make drugs that fight viruses. The body naturally produces molecules, called interferons, that interfere with viruses by attacking them. Through cloning, researchers can produce large amounts of more than a dozen kinds of interferons to use in treatment and research. Cloned interferons are used to retard the growth of certain kinds of cancer, and they can defend against viral infections, such as herpes, shingles, warts, and the common cold.

In addition to the creation of new drugs, genetic engineering has revolutionized the production and development of vaccines. Genetically engineered vaccines are purer and have the potential to prevent a greater variety of diseases than the ones produced in a more conventional manner. Vaccines introduce into the body enough of a disease-causing organism to activate the body's immune system. They work by allowing the body to form protective antibodies without ever having the disease. The first genetically engineered vaccine, developed in 1986, was a vaccine to prevent hepatitis B, a disease caused by a virus that attacks the liver. Since then, genetically engineered vaccines have been developed, or are nearly developed, for malaria, cholera, salmonella, typhoid, dysentery, and other intestinal diseases. Research is under way to develop vaccines for tuberculosis, leprosy, syphilis, herpes, and another form of hepatitis.

Gene therapy

Scientists who identify defective genes and develop diagnostic tools are working toward a single goal—the cure of genetic diseases by gene therapy. Gene therapy is the substitution of a healthy gene for an existing defective one. Several strategies are possible: gene insertion, gene modification, and gene therapy. With gene insertion, doctors insert copies of a normal gene into the chromosome of a diseased cell, and the normal gene overcomes the defective one. With gene modification, doctors alter the defective gene right in the living cell by recoding its faulty sequence of nucleotides to that of a normal sequence; they do this procedure without disrupting the layout of the gene. With gene therapy (also referred to as gene surgery), doctors remove a defective gene and replace it with a cloned, normal substitute. Geneticists dream of the day when they can accomplish lifelong cures for a host of diseases with these strategies.

Because gene therapy is difficult and has many steps, real progress is slow. After identifying and isolating a gene, scientists clone it so that many copies are available. In order to insert normal genes into the defective cell, scientists must have a suitable vector, or carrier. Then the altered genes are reinserted into the patient where they should produce new cells to replace the defective

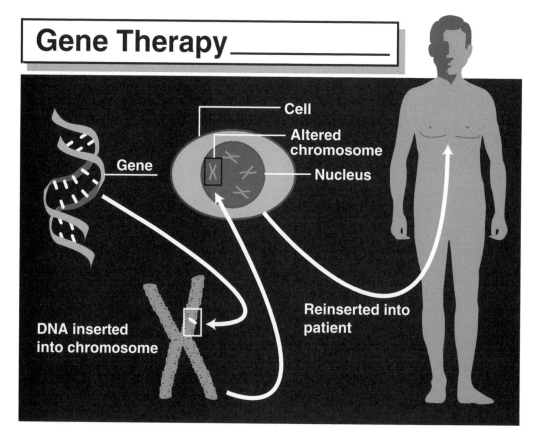

Gene Therapy

Cell

Altered chromosome

Gene

Nucleus

DNA inserted into chromosome

Reinserted into patient

Cynthia Cutshall plays catch after becoming one of the first people in the world to benefit from gene therapy, the technique of treating disease by transplanting genes.

ones. Each of these steps is difficult, and there are many risks. Because scientists cannot always find a way to insert the altered gene into an exact position in the chromosome, it is especially hard to get the altered gene to produce new cells after it has been reinserted. The altered gene could land in the middle of an existing gene when it is reinserted. It could produce a destructive mutation that disrupts the body's genetic process, or the inserted gene could cause sudden changes in neighboring genes and turn the normal growth pattern of a cell into a cancerous growth pattern. To ensure that procedures are safe, scientists do repeated tests on animals before they even apply for permission to do tests using human patients. Strict government guidelines ensure patients that tests will be reasonably safe and effective and that their rights are protected.

A few cases of gene therapy have been accomplished successfully. Cynthia Cutshall and

Ashanthi DeSilva (left) is able to attend school since undergoing gene therapy to treat a rare genetic disorder that had previously forced her to live in isolation.

Ashanthi DeSilva had a rare genetic disorder called ADA deficiency, a disease of the immune system that prevents the body from fighting infections, such as sinus infections and pneumonia. A child with ADA deficiency must live in isolation and avoid contact with infectious bacteria and viruses. In 1990 Dr. R. Michael Blaese, Dr. Kenneth Culver, and Dr. W. French Anderson removed white blood cells from the bodies of the two girls, replaced defective genes with healthy ones, and infused the blood cells back into the girls' bloodstreams. After several treatments, both girls have functioning immune systems and can attend school and participate in sports. A similar success occurred in 1993 when Dr. Donald B.

Kohn and a team of doctors used the same procedure to correct the genes of three infants with ADA deficiency. Because the patients were infants, doctors had a better chance of altering their stem cells, which divide rapidly right after birth. Genes can be engineered only during cell division. Since stem cells control production of all blood cells, doctors theorized that they could correct ADA permanently if they could alter the stem cells. All three of these patients, now two years old, are healthy and living normally.

In 1992 Dr. James M. Wilson used gene therapy to treat a twenty-nine-year-old Canadian woman who suffered from a rare liver disease that causes abnormally high cholesterol levels. She had had a

In 1990 W. French Anderson (left), Kenneth Culver (center), and R. Michael Blaese (right) made medical history by using gene therapy to repair the immune systems of two young patients.

heart attack at age sixteen and had undergone bypass surgery at age twenty-six. Dr. Wilson removed about 15 percent of her liver, separated individual cells, inserted healthy genes, and infused the engineered cells into a vein that empties into the liver. Eight months later the woman's cholesterol level had dropped 40 percent. Dr. Wilson hoped to treat her with conventional drugs to further lower her cholesterol level to normal range. Her condition has remained steady.

In addition to these successful cases of gene therapy, researchers are working on other projects. Dr. Albert B. Deisseroth and his team at the University of Texas M. D. Anderson Cancer Center pioneered the treatment of ovarian cancer with especially strong doses of a drug (Taxol) in a patient named Earlene. Because heavy amounts of Taxol damage bone marrow, the team first genetically altered the patient's bone marrow so that it would not be damaged by the drug. In November 1994, they removed portions of bone marrow so that they could alter the patient's genes. In January 1995, the doctors transplanted the genetically altered bone marrow back into Earlene's bones and made plans to administer the first of twelve strong doses of Taxol. Four months and two Taxol treatments later the patient is recovering faster than expected. The April 2, 1995, issue of the *Houston Chronicle* quotes Dr. Deisseroth: "The tumors are responding. We do not see any Taxol side effects."

Future plans for gene therapy

In another experiment at Johns Hopkins University, Dr. John T. Isaacs and a team of researchers isolated a gene that seems to prevent prostate cancer cells from moving from their original site and spreading to other organs. They have inserted this gene into human prostate can-

Albert B. Deisseroth genetically engineered a cancer patient's bone marrow so that it would not be damaged by high doses of a cancer-treating drug.

cer cells removed from patients. The next step is to insert the engineered prostate cancer cells into mice to see if they block the spread of cancer. They have many rounds of tests to do on mice before human patients can receive the altered genes.

At the Texas Heart Institute, Dr. James Willerson and colleagues are trying to insert genes into cells of the artery wall to head off blood clots and prevent heart attacks. Researchers are currently testing the technique on pigs and say it will not be used with humans for several years. At the Baylor College of Medicine, Dr. Savio L. C. Woo and Dr. Robert Grossman want to develop a gene to kill glioma, a type of deadly brain tumor. Surgery is usually ineffective with brain tumors because most of them grow like spider webs into the folds of the brain and cannot be reached. Woo and Grossman's plan is to insert a gene into the cells of the glioma. This gene would cause the tumor cells to make an enzyme that attracts a drug designed to kill the tumor. Tests on rats showed that the gene did attract the drug, and the tumor shrank. Tests on humans are just beginning.

Concerns about genetic engineering and medicine

While the possibilities are many and the hopes are high that genetic engineering will revolutionize medicine, many questions and problems remain. At the present time, few diseases have been successfully treated with gene therapy. Of the four to five thousand genetic disorders already located, only a tiny portion have cures. Because finding a cure for one disorder is difficult, costly, and time-consuming, cures for the many yet unsolved disorders lag far behind the technology to screen for them.

Screening is the aspect of genetic research in medicine that has raised the most problems and concerns. As soon as scientists identify a gene causing a genetic disorder, they can screen a patient's genome to see if the mutated gene exists. This technology raises ethical concerns about individual rights. Problems already exist, but many more will arise as knowledge about the genome increases. Scientists want to think about the problems and try to prevent them. The most serious question about screening is its accuracy. If the results of screening tests are to be the basis of significant decisions, the tests and procedures for obtaining them must be correct. Regulations should be in place to assure the accuracy of genetic screening. In addition to questions about the accuracy of screening tests, many questions arise about the way the tests might be used.

Ethical concerns

One possible use of screening concerns a couple's right to have a child. For example, suppose Sue and Bob come from families in which some members have had genetic illnesses. Both of them are healthy, but they wonder if either of them carries a gene for an illness that can be passed on to a child. They decide to be screened because they think the results will help them make an informed decision about their family. The results indicate that they have between a 60 and 90 percent chance of producing a child that will have the gene that causes a genetic disease. They must decide whether to risk having a child who may inherit a genetic disease, to adopt a child instead, or to remain childless. If they know the risks and have a child that does have a genetic disease requiring costly treatment, who will pay for the treatment? Health-insurance providers, for example, may think they should not pay because the problem could have been prevented; they could simply drop Sue and Bob's insurance. In the past, these problems did not exist because a couple had no way of knowing their risk of producing a child with a defective gene.

Another question about screening concerns the wisdom of telling the patients the results of screening. For example, suppose Arthur and Jane have two children, Ann, age ten, and Mary, age eight. Jane comes from a family in which members have had Huntington's disease, a fatal disease that begins to destroy the nervous system when an individual reaches middle age. Thirty-four-year-old Jane is worried that she may have the gene that causes the disease and she also knows that a screening test can determine whether she actually has the gene. Jane believes screening has two possible advantages: She may find out she does not have the gene and gain peace of mind, or she may find out that she does have the gene and take the opportunity to do some things that are important to her before the onset of the disease. On the other hand, Jane believes there are advantages to not being tested: She can continue to live her life as if she will remain healthy; finding out for sure that she has the gene might turn out to be an overwhelming psychological burden. Furthermore, if she does not have the test, the family's health insurance company has no reason to try to drop the family's health insurance. Jane and Arthur also wonder if their daughters carry the gene and if screening is right or wrong for them.

Facing tough decisions

The decision to screen for genetic disease is very complex. As more diseases are found to have genetic causes or components and more screening tests become available, more and more people will be faced with dilemmas such as the ones facing Sue and Bob or Jane and Arthur.

A doctor and a genetic counselor work together to help a couple understand the advantages and disadvantages of genetic screening.

Genetic counselors are professionals who help people understand the personal impact of genetic screening. They explain the meaning of test results so that individuals know both the advantages and disadvantages of being screened and, if screened, learning the results. In addition, genetic counselors offer support to those who discover that they have a mutated gene that will cause a future illness. Unfortunately, the supply of genetic counselors is too small to meet the demand. The discovery of new genetic diseases and methods to screen for them is proceeding faster than the recruitment and training of skilled counselors.

Many health care workers worry that the results of genetic tests may be used inappropriately by health-insurance companies and employers. Health-insurance companies, for example, might require genetic screening before accepting an applicant. Some legal scholars believe that requiring an insurance applicant to undergo a genetic screening test is a violation of the individual's right to privacy. Alternately, insurance companies could set high prices for people with defective genes or deny insurance altogether to people with an increased risk of developing a particular genetic disease.

Genetic discrimination in the workplace

The same potential for violation of individual rights exists in the workplace if employers have free access to screening devices. Employers could make screening mandatory and then refuse to hire anyone who has defective or at-risk genes. The worst possible scenario could be the emergence of a genetic underclass. Concerned people believe that public discussion needs to take place now to prevent genetic discrimination in the future. According to the March 27, 1994, issue of the *Los Angeles Times*, these problems already exist. Author Sheryl Stolberg writes in "Insurance Falls Prey to Genetic Bias":

> Experts estimate that either as a result of sophisticated new genetic tests or a family history of inherited illness, thousands of Americans are being discriminated against because of something over which they have no control: their genes.
>
> Many insurers, observers say, do not want to take the risk of covering someone who may require expensive medical treatment and are laying bare their genetic biases. So too are employers and adoption agencies. All of this is occurring in a legislative vacuum; because the issue is so fresh, only a few states have laws barring genetic discrimination.

What laws, if any, should govern genetic engineering and testing are currently being debated. In the future, such battles will prove even more contentious.

4

Genetic Engineering of Animals for Human Medicine

(Opposite page) Animals currently play an important role in the development of new medicines for human patients. In genetic engineering, researchers depend on the use of animals to test and produce drugs. Some question, however, whether humans have the right to experiment on animals for their own gain.

THE USE OF ANIMALS in research is what makes medical breakthroughs in genetic engineering possible. Animals are genetically engineered to manufacture medicines to treat and vaccines to prevent human illnesses. Animals are genetically engineered to test drugs and treatments for human diseases, particularly genetic diseases. Some researchers are exploring the possibility of using animal organs for human transplants. Many people voice strong emotional responses to the use of animals in medical research. Concern about the use of animals in medical research is not new, but animal-rights activists think that genetic engineering causes additional extreme suffering and new abuses. Furthermore, those who value secure, traditional views of the relationship between humans and animals are upset by the blurring that occurs when they visualize animal organs in human bodies.

65

In the early 1980s, researchers developed molecular pharming, or research in which animals are used to produce drugs that treat human diseases. In the first stage, researchers engineered human insulin, a biochemical protein, into the genes of pigs and engineered human growth hormone, a different biochemical protein, into the genes of sheep. The human genes became a permanent part of the animals' genetic codes. When the animals had created enough of the human biochemicals, they were slaughtered and the biochemicals were extracted. Researchers immediately sought ways to produce the human biochemicals without destroying the animals. By the fall of 1991, three independent research teams had successfully engineered animals to secrete human biochemicals into their milk. That advancement meant that sheep, goats, and cows could be genetically engineered to produce human pharmaceuticals in their milk; the supply was renewable, and animals did not have to be killed.

Genetically engineered drugs

Today these so-called animal factories produce many pharmaceuticals that are sold commercially. The first genetically engineered drug available for commercial production in 1983 was Humulin, or human insulin, a protein that regulates glucose, a form of sugar found in the blood, for people with diabetes. The second was Protropen, a human growth hormone used to treat dwarfism, a condition that causes children to be undersized because they have an inadequate supply of their own growth hormone. Doctors also use Protropen to accelerate the growth of new tissue in patients who suffer from burns and injuries. Animal factories also produce medicine to dissolve blood clots in heart patients and blood-clotting drugs for treating hemophilia, a failure of the blood to coagulate. Molecular biologist Aya Jakobovits has engineered a mouse with a complete range of human antibodies, the proteins that protect the body from invasion by disease organisms. Antibodies produced in mouse factories are used in medicines and in other experiments.

The role of animals in studying human diseases

Although a variety of animals are used in research, mice are used more than all others for two reasons. The genetic system—the genome—of mice, although not identical to the human genome, is similar enough so that reactions in mice resemble reactions in humans. In addition, mice reproduce quickly, and researchers can study several generations of mice in a short time. Animals are essential to the study of genetic diseases because initial experiments searching for a cure for a genetic disease cannot be done on humans. Injecting humans with genes to cause a disease is illegal, and even if it were legal, studying a disease over a human lifetime takes too long. Researchers must try many remedies on sick animals and find one that repeatedly and safely cures the animals. Then they can begin trials on human patients but only with the consent of the patient.

To study human diseases with genetic components, scientists engineer human genes that cause the disease they are studying and experiment with grafting, cloning, or altering the genes of the research animal. The mutated gene for any disease that has been located in the human genome can be removed and engineered into an animal's genome. For example, mice at the University of North Carolina have the mutated gene that causes cystic fibrosis. Mice at the University of Minnesota have been injected with leukemia cells, a genetically related cancer of the bone marrow.

A laboratory mouse involved in an Alzheimer's disease study receives an injection. Because their genetic system is similar to that of humans and because they reproduce quickly, mice are often used in genetic engineering research.

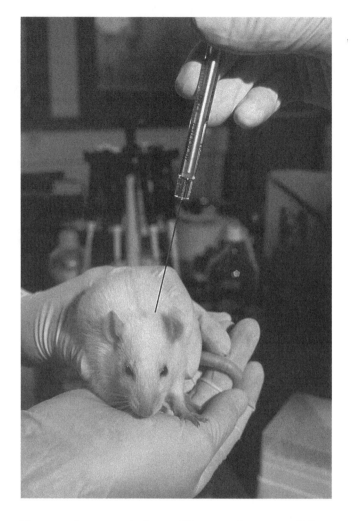

Researchers at Athena Neurosciences of San Francisco and at the Eli Lilly Company in Indianapolis, Indiana, report that they have successfully inserted human Alzheimer's disease genes into mice and are waiting to see if the mice develop the disease. Some laboratory mice have Lou Gehrig's disease. Other laboratory mice have AIDS. Rats and pigs with clogged arteries, a condition that causes heart attacks in humans, live in laboratories at the University of Chicago.

Scientists study these sick animals in an attempt to recognize the causes of disease and to test potential cures. For example, Dr. F. M. Uckun of the University of Minnesota tested an antibody that attacks a molecule on the surface of leukemia cells. Ten mice with leukemia received the treatment, and 110 mice with leukemia received no treatment. The treated mice lived twice as long as the untreated mice, and more than 99 percent of their leukemia cells were killed. The 110 untreated mice died within sixty-one days. Researchers study the mice with Alzheimer's disease to try to discover how and why plaque, which is composed of Brillo-like balls of protein, develops in the brains of patients with Alzheimer's disease. Vice president for research Dr. Ivan Lieberburg, at Athena Neurosciences in San Francisco, devised complex tests to monitor the decline of memory in the sick mice. This project was in progress at the time of this writing and conclusive results have not yet been reported. Scientists inserted small, gene-coated balloons through the clogged arteries of the rats and pigs at the University of Chicago. The internal walls of the animals' arteries absorbed the genetically engineered substance from the coating on the balloons. Running a balloon through clogged human arteries to open them is a treatment called balloon angioplasty, but the arteries often close again soon after the procedure. After the animal arteries had absorbed the engineered coating, between 50 and 60 percent of the rats' and pigs' arteries remained open.

Sick animals for sale

The large numbers of research projects that use mice have generated a new industry. In 1991 commercial laboratories began to supply researchers with mice already engineered with a disease. Filling custom orders for rodents has become routine in many laboratories. David Winter, president of GenPharm, calls the business "dial-a-mouse." To prepare a mouse for sale, a scientist locates the human gene that causes a disease and inserts it into cells taken from a mouse embryo. After cloning, a technician locates the mutated cell from the gene library and uses a thin, sharp needle to push it into another mouse embryo. A trained technician can inject mutated cells into about two hundred mouse embryos in an afternoon. In three weeks mice are born with the engineered human disease. Purchasing mice already engineered with a particular disease is expensive, but the practice saves time for researchers who are trying to find cures for human diseases.

Patented animals

The demand for research mice already engineered with a disease has led to the practice of patenting them. A patent protects the exclusive rights to create, use, and market a creation for seventeen years. In 1980, the Supreme Court decided in a five-to-four decision that living organisms could be patented. In 1987 the Reagan administration ruled that genetically engineered or otherwise altered animals could be patented. With a patent, the research team that develops the techniques to engineer a specialized mouse can control the model and make a profit on sales. With the lure of patent profits, many researchers and corporations have begun to engineer a wide range of research and farm animals. The first patented animal was the Harvard University OncoMouse engineered to be susceptible to cancer. Since 1987 the U.S. Patent and Trademark Office (PTO) has issued additional patents on genetically engineered animals, and more than two hundred requests are patent pending; that is, waiting for the PTO to act on the patent application. Recently scientists have considered the idea of applying for a copyright on genetically engineered products instead of a patent because a copyright has a longer lifetime. With a copyright, a scientist who creates a new genetically engineered cell or creature has the exclusive right to profit from it until his or her death and fifty years beyond.

Animals engineered for human organ transplants

In the early 1990s, scientists began research to use pig organs to fill the growing demand for human transplant organs. To accomplish this goal, scientists need to engineer the genes that control inheritance, rather than genes that control only the body of a single individual.

Cells that make up the body of an individual animal or person are called somatic cells. If the genes in these cells are engineered, the outcome affects only one specific individual. For example, if the genes in the somatic cells of a mouse embryo are engineered to produce big ears, the newborn mouse will have big ears, but its offspring will have normal-size ears. The cells that pass on traits to offspring are called germ-line cells. If genes in the germ-line cells are engineered, the outcome affects the next generation and all the generations that follow. Scientists have the technology to alter the germ-line cells of fertilized eggs.

Using this technology, two scientists began a project to engineer pigs to be factories for human transplant organs. In 1992 John Atkinson of the Washington University of Medicine in St. Louis, Missouri, and David White of Cambridge University in England engineered pigs with human proteins to prevent their immune systems from attacking a foreign substance. Atkinson and White engineered germ-line cells to supply themselves with new generations of research pigs. To

John Atkinson of the Washington University of Medicine engineered pigs with human genes as part of an experiment to develop pig organs that can be transplanted into humans when human organs are unavailable.

accomplish this step, they removed a fertilized egg from a sow, or female pig, injected it with human genetic material, and replaced it in the sow. In December 1993, Astrid, the first pig with human genes, was born. Since then, second- and third-generation pigs with human genes have been born, and the first ones have grown into adult pigs. Atkinson and White think they have developed pig organs that will be suitable for humans, both in size and in capacity. In 1996 Atkinson and White plan to transplant pig organs into baboons, whose genome is similar to the human genome. If the experiments are successful, they hope that pig livers, hearts, kidneys, glands, and lungs can be transplanted into humans when human organs are unavailable.

Controversy over the use of animals in medical research

The use of animals in medical research has been an emotionally charged issue for many years, but genetic-engineering research has renewed the opposition of those who lobby for the protection of animals. The most extreme animal-rights activists want to stop all uses of animals in medical research. Those opposing the use of animals in medical research have three major objections. First, they think that infecting animals with human diseases is cruel treatment for animals.

People also object to the use of animals in medical research on moral grounds. They believe it is immoral to use animals to manufacture human medicine or organs for human transplants, especially if the goal is to make a profit. Using pig organs for human transplants is doubly objectionable to some people. Their religious beliefs clearly distinguish between humans and animals, and implanting animal organs in humans blurs that distinction. They believe it is wrong to give human genes to a pig and unthinkable to put a pig's liver in a human body. To the most strident opponents, any alteration of an animal's genetic code is an assault on the animal's dignity and biological integrity.

Advocates for the use of animals in medical research may be less outspoken, but they strongly believe that animal research is a necessity. They acknowledge that animals experience distress in research projects, but animal welfare laws have done much to protect animals. The first animal protection law, called the Laboratory Animal Welfare Act, was passed in 1967. This law governs cleanliness and other laboratory conditions for dogs and cats. The law has been amended many times. One amendment states that before scientists can receive federal research funds, universities and corporations must review each project and weigh its potential benefits for medicine against the amount of distress it will cause the an-

imals. If the animal distress is too great compared to the potential benefit, the grant is withheld. Another amendment requires that laboratory dogs be exercised daily and that psychological care be provided for primates. Complying with the last amendment has added millions of dollars to the cost of research.

The strongest argument in favor of using animals in medical research is the beneficial knowledge obtained from animal experiments. Furthermore, researchers say animal testing is necessary to prove that drugs and treatments are safe for humans. As one scientist asked, "Where does society expect scientists to find cures for AIDS and other diseases without facilities [animals]?"

5

Genetic Engineering and Industry

BEFORE 1975 scientists in university laboratories conducted most of the basic genetic-engineering research, and the government funded their research with grants. In 1975 the Cetus Corporation became the pioneer in the use of genetic engineering for commercial purposes. The next year, another corporation, Genentech, announced plans to use hormone research to develop medical products for sale. Ten years later, dozens of large companies had their own research and development branches to develop genetically engineered products for the marketplace. Between 1979 and 1987, 250 companies began using genetic engineering to make commercial products. By the year 2000, sales of genetically engineered products are expected to be in the tens of billions of dollars.

Industry's accomplishments and problems

Industry has already marketed a number of genetically engineered agricultural products. Companies produce cloned plants, seed for genetically engineered rice, and many varieties of disease-

(Opposite page) The use of genetic engineering in industry has grown rapidly within the past decade. Hundreds of companies now use genetic engineering to make commercial products.

resistant plants and seeds. In Scandinavia, for example, microalgae are used to engineer vaccines for commercial fisheries. In 1990 only 5 percent of all trout and salmon produced in fisheries were vaccinated; today nearly all are.

Industry is also using genetic engineering to produce foodstuffs. In May 1992 the U.S. government authorized the sale of genetically engineered foodstuffs without requiring the products to carry special labeling. Since then microalgae have been used to produce vitamins, and enzymes have been used to produce cheese and meat tenderizers. In addition to milk produced with hormones and the FLavr Savr tomato, genetic engineering technology is developing food enhancers and new food flavors. The technology can also convert ordinary plant starch into sweetener to be used in a variety of new foods. Compa-

nies are using amino acids to add nutritional supplements and flavors to fruit juices.

In addition, for-profit companies are marketing many medical products, including insulin, vaccines using yeast and animal cells, and a number of pharmaceuticals from genetically engineered hormones. Industry also spends time developing diagnostic tests. Every time a gene for a new disease is discovered, competition accelerates to be the first company to produce corresponding screening devices.

Insuring genetic tests

The insurance industry also becomes involved in genetic-engineering advances. Because these tests are expensive, insurance companies have to decide which tests they will cover. On March 28, 1994, the *Los Angeles Times* described the dilemma for insurance companies this way:

> Already, companies are facing pressures to pay for unnecessary genetic tests. And Kaiser [Permanente in Northern California] and other companies are funding significant numbers of so-called "anxiety amnios"—expensive amniocentesis tests for pregnant women under 35 who are at no particular risk of having a child with a birth defect but are apprehensive nonetheless.

> New genetic advances have the potential of being a "black hole" for soaking up much larger portions of an insurer's funds, said Dr. Edward Schoen, a Kaiser pediatrician. Every time a new gene is discovered, such as those that cause cystic fibrosis and colon cancer, it opens the possibility of widespread—and expensive—testing for its disease.

Fierce competition has arisen among companies that want to patent animals altered with foreign genes, especially transgenic research animals engineered with human diseases, and sell them to research scientists. Private companies and university researchers say that they cannot spend millions of dollars engineering and breeding animals for other scientists without protecting their own investments with patents. After the first patent was granted to Harvard University for OncoMouse in 1988, companies have tried to team up with university scientists to obtain the exclusive rights to sell transgenic research animals. The commercial businesses are in the middle. They neither create nor use the animals; they make their profits by distributing the genetically engineered animals. Prestigious universities, such as Massachusetts Institute of Technology (MIT), Harvard, and Columbia, have signed contracts that give commercial companies the rights to market university-developed gene-altered animals. In an article in the May 9, 1993, edition of the *Los Angeles Times*, Lita Nelsen, licensing director at MIT, says, "We're not going to allow rapacious pricing or restrictions. But the fact that it is a profit-making entity does not seem to us to be against the American way." Lita Nelsen created a mouse with an altered immune system that is sold exclusively by GenPharm.

There are arguments for and against the for-profit sale of gene-altered animals. On one hand, researchers deserve to be paid for their investment and the creative know-how to make an animal research product. On the other hand, nonprofit researchers are unable to purchase engineered animals for their research if profit-hungry companies charge exorbitant prices.

Some risks involved

Many genetically engineered products have not been on the market long enough for scientists and society to assess their full or long-term effects, even though testing standards are high. For example, a carefully tested amino acid used as a food supplement had adverse effects. Liebe F. Cavalieri reported in the December 1995 issue of the *World & I:*

> In 1989, the *New England Journal of Medicine* reported illnesses and deaths caused by an amino acid, tryptophan, that was used as a food supplement. Eventually, some five thousand people became ill and thirty died. The product had been made by the Japanese firm Showa Denko, using genetic engineering techniques. Analysis revealed the product to be 99.6 percent pure, the impurity being a dimer, that is, two molecules of tryptophan joined together. No one had anticipated that a minor impurity might be toxic.

Nevertheless, scientists and businesspeople are predicting that genes will become the "technological slaves" of industry. The pace of development picks up every year. As G. J. V. Nossal says in *Reshaping Life: Key Issues in Genetic Engineering:* "The pathway from research bench to mass markets is thorny and tortuous, particularly where legitimate environmental and regulatory concerns enter the picture." If environmental concerns are absolved and regulations are relaxed enough for companies to produce a profitable product, the pace of change will continue to quicken.

Industry's beneficial uses of microorganisms

Though microorganisms such as bacteria and microbes have been used from the beginning of genetic-engineering research and development, industries are rapidly developing new ways to use the genetically altered microorganisms for environmental and industrial purposes.

Industry is working on many ideas and plans for the use of microbes. Already, engineered organisms remove oil spills from water and from beaches. In 1980 the U.S. PTO awarded a patent for a product that had three genes engineered into one bacterium that demonstrated a voracious appetite for oil. Other engineered microbes could

Scientists predict that genetically altered microorganisms will one day convert garbage into energy.

convert America's yearly pile of garbage into energy. Garbage, however, is so varied that it would take a very large number of microbes to break it down, a feature that makes the process expensive to research and develop. In addition, researchers are engineering microorganisms that will consume animal waste from livestock feedlots, digest it, and turn it into a gas that can be used for energy. Although the idea of converting garbage and animal waste into energy sounds good, especially to environmentalists, industry looks at costs. Profit-motivated businesses are likely to wait until energy is no longer plentiful and cheap before investing their resources in learning how to turn garbage and waste into profitable energy in a safe way.

Putting microorganisms to work in the mining industry

Microorganisms have a potentially important role to play in the future of metal and mining industries. Some microbes love metals. They will leach metals from rocks and low-grade ores. For example, microbes recover uranium and copper from solutions drained from underground mines. Already 10 percent of U.S. metal production depends on bacteria that liquefy metals. Microbes can extract silver from the waste silver sulfide solutions used at Kodak. Because microbes tolerate arsenic, a highly poisonous metallic substance present in gold mines, they can help to mine gold. At the present time, using genetically engineered microorganisms to recover metal is not cost-effective because the earth still has a supply of easy-to-recover ores and companies can still make a profit extracting them.

Microorganisms have other uses in waste management and water treatment. They can extract toxic materials from waste before factories release their waste into rivers and waste-management sites. For example, bacteria can remove about 90 percent of the lead, zinc, nickel, and cadmium from industrial wastes. Some bacteria can survive and grow in temperatures near boiling. These bacteria are used in heated industrial wastewater to cleanse it of toxic metals. Water-treatment plants, on the other hand, need a variety of engineered microbes that can attack the varied mess that runs into sewer pipes: industrial wastes, street and parking-lot runoff, solid chunks of sewage, slime, and grease. Researchers are already engineering microbes that have enhanced degrading properties. Again, cost is a problem. Large amounts of money will have to go into research and development before microorganisms will keep the environment clean and make water pure. If microbes are proven to be safe for use in public utility projects like these, industry may need the help of government to fund them because the government is the appropriate source for funds to accomplish helpful and needed projects that cost too much for profit-making companies.

Destructive uses of microorganisms

Though thousands of researchers around the world conduct experiments to engineer microorganisms for the benefit of society, it is also possible to engineer microorganisms for destructive purposes. Every genetic technique developed for helpful medical therapy can also be used for harmful military purposes. Genetic engineering can make very sophisticated biological and chemical warfare weapons. Many governments around the world have military programs to develop biological weapons, some of which are developed by industries that have contracted to do research and development. According to David Suzuki and Peter Knudtson, the U.S. military budget for biological weapons increased from $15 million in 1980 to $39 million in 1985. The military believes it must develop an arsenal of defensive biological weapons to use against enemy attack.

There are several ways that genetic engineering can be used to make efficient biological weapons. Specialized microbes have been engineered into efficient killers by making them hardier, longer lasting, and more resistant to antibiotics. The destructive microbes can be mass produced by cloning and stored for later use. Even more sinisterly, *Military Review* reported that the U.S. military has the technology to develop "ethnic weapons." Since certain ethnic and racial groups are vulnerable to particular diseases, scientists have the tools to engineer microbes to target special groups.

Another biological weapon could be manufactured by cloning a biological poison inside bacteria or yeast host cells. Huge quantities of weapons filled with snake venom, shellfish toxin, or bacterial toxin could be produced quickly and cheaply. Perhaps the most powerful and efficient weapon could be made by using botulin—the toxin excreted by the microbe causing botulism, a virulent food poisoning. One-millionth of a gram of botulin will kill a human. Less than two pounds of botulin in the water supply for a town of fifty thousand people would kill 60 percent of the people in twenty-four hours.

Other biological weapons can attack a country's agriculture. In Third World countries where agricultural diseases are often poorly monitored, intentional spreading of a disease may look like a natural epidemic. Animals could be attacked through the food supply. Crop seeds engineered with harmful toxins could be sold to an enemy nation. Infected seeds are especially easy to use as a weapon if one crop predominates in the area.

Advantages and disadvantages of biological weapons

Biological weapons have many advantages. Biological weapons can be more selective in targeting certain populations. They can weaken or destroy troops without destroying property. Another advantage is their variety. Because there are so many human and agricultural diseases, designers of biological weapons have an abundance from which to choose. Moreover, biological weapons made by cloning are cheap. Finally, they are suitable for secret attacks. They can be engi-

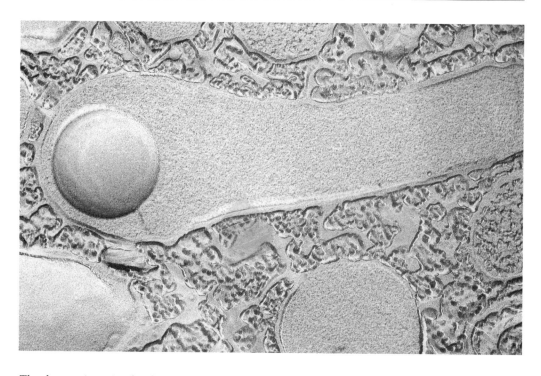

The destructive microbe that causes botulism could be used to make powerful and deadly biological weapons. Just one-millionth of a gram of the toxin will kill a human.

neered to activate when no troops are present. They are small and can be "fired" through a number of delivery systems.

Most people and governments around the world, however, are morally opposed to the use of disease as a weapon. Any country that uses such weapons is at risk of censure from other countries. At the 1972 Biological Weapons Convention, about half of the nations of the world, including the United States, signed a treaty pledging not to produce biological weapons and to disarm the ones already produced. Though scientists see the treaty as a start, many feel it is weak because it allows research and the development of defensive biological weapons, which are essentially little different from offensive weapons.

Industries in a global economy

In the growing global economy, both competition and control are important. Because the

United States has a well-developed scientific base and an abundance of highly qualified scientists, U.S. industries lead the world in the use of genetic engineering to develop products. However, because American people demand safety and security, testing to meet high standards drives up costs for U.S. firms. Recently, U.S. corporations have begun to develop partnerships with businesses in other countries. Partnerships and competition in a global economy increase the danger that companies could be driven to produce harmful products for profit or to circumvent testing and high safety standards to cut costs. For this new and expanding technology, policies and regulations in the United States are still being debated, and debate on international policies and regulations lags behind the national debate.

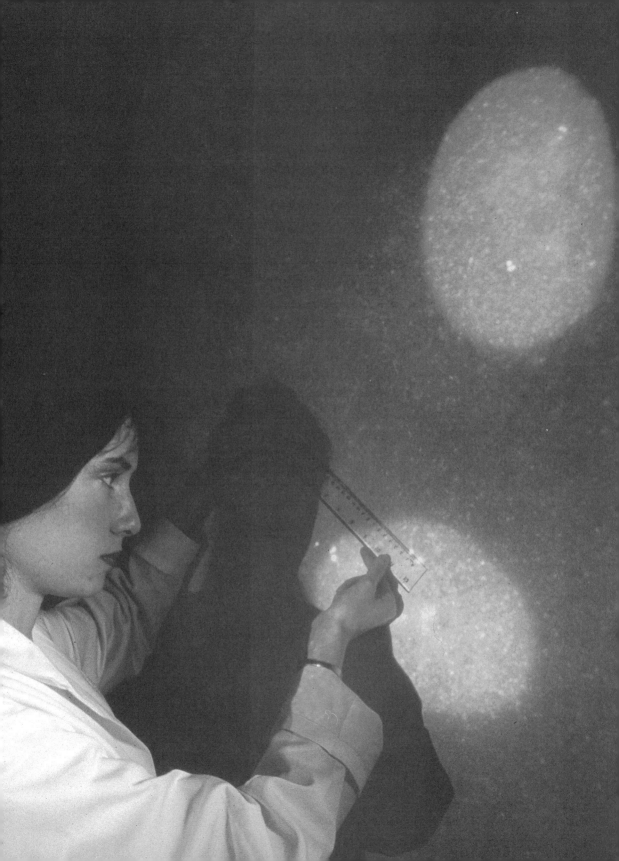

6

Genetic Engineering and the Future

THE FUTURE OF genetic engineering depends on the integrity and creative imagination of scientists, governments, industries, and the public. There will be good judgments and perhaps some poor ones. Good but imperfect outcomes are the nature of all complex human endeavors.

The Human Genome Project

The possibilities envisioned with genetic engineering have evoked many predictions. One prediction already under way and likely to be completed is the Human Genome Project. The goal of the project is to produce a map of all of the estimated 100,000 human genes. Mapping involves identifying and recording the location of specific genes. When scientists finish, they will have identified the chemical sequence in approximately three billion base pairs of human DNA.

Such a huge undertaking involves scientists around the globe. In America, the National Institutes of Health is organizing the work, and the government is funding it. The government allocated more than $300 million the first year and

(Opposite page) As part of research for the Human Genome Project, a scientist studies a human cell nucleus that has been projected onto a screen to allow measurement. The genome project is the foundation of the future of genetic engineering in medicine.

approximately $20 million more over the next two years. Governments in Canada, England, France, Italy, Australia, and Japan have also provided funds for their scientists to work on the mapping project. Interestingly, there is no overall central plan, but the most recent interim report stated that the separate projects around the globe are falling into place and forming a pattern. Perhaps the success of the piecemeal approach can be attributed to the expert biologists and geneticists from around the world who form the Human Genome Organization, whose goal is to coordinate the findings from all of the countries and to manage the information. Scientists began the pro-

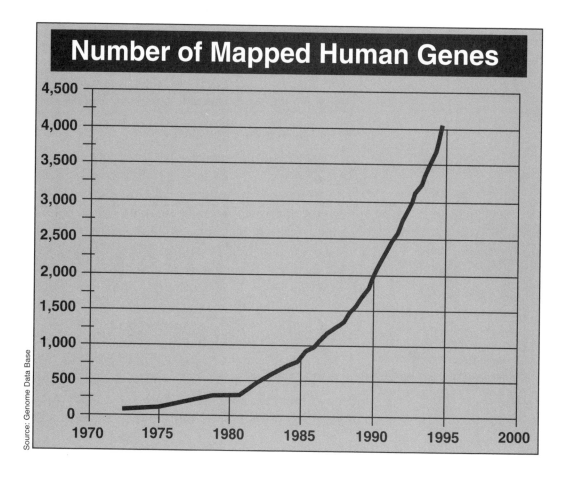

ject in 1986 and hope to complete the map by the year 2005, but the project is moving faster than expected.

"One of the most powerful tools"

Scientists hope that the genome project will help improve human health. The ultimate goal of doctors is to prevent and cure human diseases, but that goal involves many small steps. The complete map will allow scientists to design screening tests for any of the genes identified on the map. With a completed genome map and new screening devices, doctors will be able to diagnose illnesses by comparing the genomes of healthy and sick people. Eventually, with a full array of screening tests, they hope to detect defective genes, engineer them, and prevent illnesses. A government publication, *Mapping Our Genes: Genome Projects: How Big, How Fast?*, states that the map of the human genome "will provide one of the most powerful tools humankind has ever had for deciphering the mysteries of its own existence."

While most researchers in the Human Genome Project identify genes, some devote their time to organizing the information and storing it on computers; they update the databases every forty-eight hours. Other members of the mapping team work on communication. They produce books about the genome and newsletters about the latest findings. They produce catalogs of available materials and maps of small sections of the genome. Recently a Minnesota company put the project information on the Internet. When mapping is complete, there will be public databases for the complete human genome. "The ease with which researchers can retrieve and use the data from these and related databases will provide one measure of the project's success," according to a

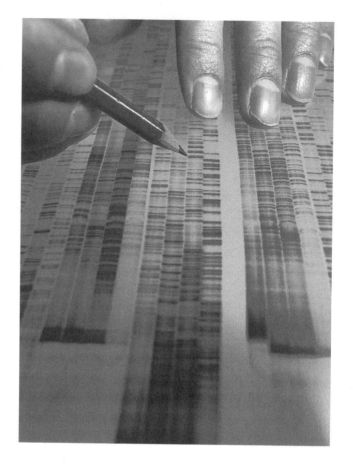

A scientist examines DNA sequencing. The goal of the Human Genome Project is to decode the exact sequence of the nearly three billion base pairs of human DNA.

government newsletter *Human Genome News.* The genome information center distributes progress reports to genetic scientists, school science teachers, and ordinary interested citizens.

The mapping of the human genome may lead to breakthroughs in drug treatment and gene therapy. Scientists have already found a drug to attack leukemia, and it is probable that they will find treatments for other cancers as the genome map nears completion. If gene therapy projects, like the ovarian cancer project Dr. Albert B. Deisseroth is conducting in Texas, turn out to be successful, it is probable more projects will follow. As the Human Genome Project nears completion, doctors and patients have greater expectations that defective genes can be fixed. However, even if scientists can alter genes easily, the costs of gene surgery and drugs developed by genetic engineering are very high. Because the available sources of private and public health-care funds are already strained, genetically engineered medicine will either overextend the present sources of funds or health care may have to be rationed and some people will go without. Furthermore, as new technologies and therapies develop, doctors and nurses will probably need retraining, which will also require funding.

Dramatic changes in the future

Futurists predict dramatic changes in the practice of medicine. Based on the findings in the experiment with the fly eyeless, scientists hope to discover genes that control whole components such as organs and limbs and genes that control behaviors. With these findings, they hope to discover exactly what makes humans and higher primates, such as chimpanzees, different. Ninety-nine percent of the two genomes are the same, but the 1 percent that is different has significant meaning. In addition, scientists speculate that genes that control complex traits may also unravel the secrets of aging and of psychiatric disorders.

Many scientists predict that the delivery of medicine will undergo drastic changes. They foresee the day when doctors practice predictive medicine. With a full array of screening tests, doctors will screen their patients' genomes once a year. They will repair any defective genes before illness sets in. In an article entitled "Scientific Knowledge and Human Dignity," published in the September 1994 issue of the *UNESCO Courier*, French doctor Jean Dausset says:

> And prevention means predicting. . . . With predictive medicine, we shall certainly be able to avoid

much pain and suffering and perhaps even life in perfect health to a ripe old age. In short, medicine will be tailored to the individual and will prove less costly and more effective.

These are rosy predictions made at a time when gene therapy is just beginning; so far scientists have altered only a few single genes and do not yet know if altering complex genes is a possibility.

Advances in tools and techniques

Scientists and technologists are developing new and refined tools for genetic engineering. Computer software can be programmed to make many identical copies of a gene quickly and identify the sequence of chemicals in a base pair. Other new tools help scientists insert foreign ge-

Computer software designed especially for genetic engineering research speeds the process of identifying gene sequences.

netic material into genes more accurately. For example, tiny glass needles that work like the needles used for injections and another gunlike tool shoot genetic material into a specific spot in a gene. Technicians are developing carriers, the microorganisms used to clone recombined genes and make the right gene for insertion. New carriers can put foreign genetic material at a specific site in the genome and start the expression of new combined genes at a preset time. Computer software can display gene proteins in colored three-dimensional images. One researcher used rare metals to make specified molecules glow with a pink light. These new genetic-engineering technologies streamline research, make new procedures possible, and help scientists locate and manipulate genes faster and more easily.

Future projects

Given the speed and success of mapping the human genome, scientists will probably map the genomes of other organisms. The mouse genome is already mapped because mice play a vital role in medical research. Geneticists from the University of California at Berkeley, the University of Washington, and the University of Oregon are working on a map of the dog genome, which contains seventy-eight chromosomes. The researchers hope their experiments will pinpoint the genes that determine dogs' distinctly different shapes, sizes, and behaviors. Already they have found that a variation on chromosome 16 governs which dogs are barkers and which are more silent.

Gene maps of crop plants like corn, tomatoes, and cabbage already give plant breeders tools to develop new varieties. Scientists expect to use existing technology to map other species of plants and animals.

"We've done it; Rothchild! The perfect watchdog."

Probable outcomes of genetic-engineering technology are the hardest for scientists to predict. Enough research has been done to make many of these projects possible, but there are too many unknowns to predict future developments accurately. On the other hand, predictions abound concerning what the distant future will bring. Perhaps it is safer to dream about possibilities than it is to foretell what might develop. All areas of genetic technology have speculators—agriculture, medicine, and industry—and speculators have a broad vision.

Enthusiastic scientists speculate about the possibilities of plant and animal research. For example, a scientist concerned about the environment envisions cloning trees to reforest areas that have been stripped bare by logging. Another scientist predicts that fish can be engineered with antifreeze to withstand very cold ocean water. Perhaps new growth hormones will produce cows that weigh five tons and pigs that measure twelve feet long and five feet tall. One scientist foresees the day when humans have the ability to engineer animal behavior traits and could, for example, engineer a wolf to ignore sheep. In an article entitled "Facing the Future: Genetic Engineering," published in the *Animals' Agenda*, Michael Phillips of the Office of Technical Assessment said, "Right now, we don't know what the limits are. . . . All the traditional rules we thought about the animal kingdom . . . are thrown out the window."

Industry

Speculation about the future use of DNA technology also occurs in industry as in the example of a computer that does mathematical calculations based on DNA pairing. In the April 11, 1995, *New York Times* story "A Vat of DNA May Become Fast Computer of the Future," Gina Kolata reports on research conducted by computer theorist Leonard Adleman of the University of California at Los Angeles. Adleman designed a computer using DNA. Instead of representing information with O's and 1's as in conventional computers, Adleman represented information with the chemical units of DNA—A's, T's, C's, and G's. Instead of programming a conventional computer to do calculations by telling electrons to travel on particular paths, Adelman ordered units of DNA chemicals in particular sequences and combined them with DNA molecules to synthesize in a test tube. The sequence of the new molecule that formed from the synthesis was the answer to his mathematical problem. Though skeptical, computer experts are very interested in Adleman's model, some even excited.

Kolata explains the capacity of a DNA computer:

The advantages of DNA computers would be that they are a billion times as energy efficient as conventional computers. And they use just a trillionth of the space to store information. . . . It would be possible to do more operations at once than all the computers in the world working together could ever accomplish. . . . A DNA memory could hold more words than all the computer memories ever made . . . and in the bottom of a test tube could store at least a million times as much information as the brain.

Genetic engineering and problems of policy and regulation

At the present time, several government agencies regulate genetically engineered products. They are the Department of Agriculture, the Food and Drug Administration, the Environmental Protection Agency, and the Department of Labor (for safety in the workplace). The National Institutes of Health control genetic-engineering experiments. Their appointed committee regulates experiments to ensure that they are safe. Scientists, lawyers, and ethicists comprise this committee, which reviews requests for grants to do research for the Human Genome Project and for other projects.

Several levels of government control are in place. A general rule says that scientists are free to pursue their own scientific interests, but if they receive public funds, their research must respond to society's need. In addition, there are specific direct and indirect controls enforced by the government department that regulates the policies:

1. Each organization performing research must have safety rules, a professional safety officer, and a safety committee. The research must conform to the safety rules.
2. Research receiving government funding must be certified safe by the ethics committee and an appropriate specialized committee. This

rule applies to all recombinant DNA research involving the engineering of genes.

3. Researchers who conduct research on humans must obtain both informed consent from the person and the approval of the hospital's or university's ethics committee, which is to include both medical experts and laypeople.

4. Scientists are answerable to administrative controls of the institution for which they work and subject to disciplinary action if they act outside accepted norms.

Researchers are indirectly controlled by peer group pressure: Approval from their peers is very important to scientists, and they do not want to risk loss of respect by acting unethically. In addition, scientists are under pressure to publish papers in scientific journals; publications refuse manuscripts that fail to describe the safeguards

Researchers harvest laboratory-grown barley plants that have been genetically engineered to resist disease. Currently, several government agencies regulate the safety of genetically engineered products.

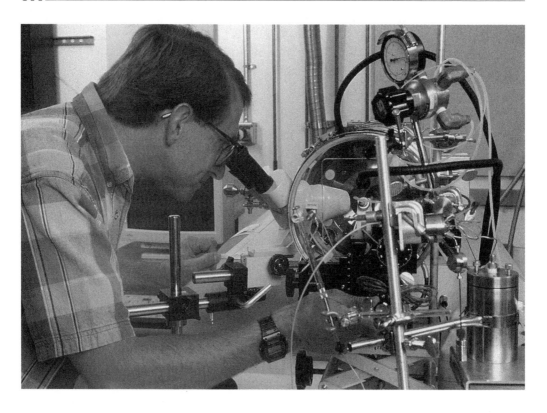

When undertaking genetic engineering research, scientists are pressured to act ethically by the government, their peers, and the institutions for which they work.

used in the research, a practice that provides another indirect control.

With the recent rapid changes in genetic engineering, regulations established in the 1980s need updating, especially some of the safeguards imposed by committees. In an article entitled "Genetics in the Marketplace," published in the September 1994 issue of the *UNESCO Courier*, Ezra N. Suleima writes: "The fact is that the issues raised by biotechnology [genetic engineering] cry out for policies that establish the contours within which scientific advances can legitimately occur." If policies and regulations need to be revised, then decisions must be made to determine who does the controlling, the kind of controls, and the degree of control.

Policy makers agree on one point—that blanket rules governing all of genetic engineering are hopelessly unworkable, since genetic engineering is a set of tools. Instead, mechanisms must be set up to review projects on a case-by-case basis to maintain control over the outcome of genetic engineering, but not the process. Policy makers also agree that government, industry, scientists and other experts, and ordinary citizens all need to share control of genetic engineering. Policy makers agree less on how much control is necessary. Advocates for less control argue that excessively tight regulations add such high costs that American companies cannot get products into the market and, therefore, cannot compete with companies from countries with less stringent laws. Furthermore, if regulations are too tight in the United States, American companies will do their testing in countries with more permissive laws. Skeptics who oppose the use of genetic engineering want strict regulation by the government with the help of environmental organizations, public-interest groups, concerned individuals, and scientists who have misgivings about genetic engineering.

The distant outcome

Whether the genetic-engineering revolution produces great benefits or great harm is a question that will not have an answer for many years. The leaders of the revolution—the scientists—can make knowledge and tools available, but they cannot assure that people will use them for the benefit of humankind. Philosopher of science and Nobel Prize–winner Sir Peter Medawar honored all scientists when he argued that "science, broadly considered, is incomparably the most successful enterprise human beings have ever engaged upon." The future will tell if the genetic-engineering revolution is one of science's most successful enterprises.

Glossary

allergens: Substances, such as pollen, that cause allergies.

antibiotics: Substances used to destroy or inhibit the growth of bacteria and other organisms.

bacterium (plural, bacteria): A single-cell microorganism; some kinds of bacteria cause diseases.

base pairs: The pairs of chemicals, attached to one another, that bond and make the rungs that connect two strands of the DNA double helix.

biochemical proteins: The chemical substances in living organisms.

bioethics: The study of the moral and ethical implications of new biological discoveries and biomedical advances.

biological warfare: The use of disease-producing microorganisms or toxic biological products to cause death or injury to humans, animals, or plants.

BST: The commercial growth hormone, called bovine somatotropin, which is fed to cows to accelerate milk production and meat growth.

cell fusion: A genetic-engineering procedure in which cells from two unlike species are merged into one.

chromosome: A threadlike strand of DNA and associated proteins in the nucleus of cells; chromosomes carry the genes and transmit hereditary information.

clone: To make multiple identical copies of a DNA sequence.

copyright: The legal right granted to the creator of an artistic or genetic work to exclusive production, sale, or distribution of the work for the creator's lifetime and fifty years beyond.

crossbreed: To produce an organism by mating individuals of different breeds.

cross-pollinate: To transfer the pollen from the flower of one plant to the flower of another plant.

defective gene: A gene with a harmful mutation.

diagnosis: The process of identifying or determining the nature and cause of a disease.

DNA: A nucleic acid that carries the genetic information in the cell and is capable of self-replication; it consists of two long chains of nucleotides twisted into a double helix and joined by hydrogen bonds.

double helix: The double strand of DNA wound into a spiral structure.

embryo: An organism in the very early stages of development before it reaches a recognizable form.

enzyme: Any of numerous proteins produced by living organisms.

eyeless: The name given to a fruit fly gene by the Swiss scientist who discovered the gene that controls development of the whole eye.

food enhancer: Any product added to food to improve its taste or nutritional value.

gene: The fundamental unit of heredity; one segment of DNA arranged in a specific sequence that is passed on from parent to offspring.

gene insertion: Inserting copies of a normal gene into the chromosome of a diseased cell, allowing the normal gene to overcome the defective one.

gene library: A collection of DNA that results when genes from two species have separated and recombined randomly.

gene modification: Alteration of a defective gene within the living cell by recoding its faulty sequence to a normal sequence.

gene therapy or surgery: Removal of a defective gene and replacing it with a cloned, normal substitute.

genetic counselor: A trained professional who works with patients who are deciding whether or not to be screened for a disease or who have learned that they have a mutated gene that causes or will cause a genetic disease.

genetic disease: Any disease caused by a defect or mutation in the genome.

genetic engineering: Techniques that change the characteristics of an organism by altering its genes.

genome: The complete genetic system of an organism.

germ line: The cells containing the genes that control the traits passed on to offspring.

hormone: A substance produced by one tissue and conveyed by the bloodstream to another to cause growth or affect metabolism.

Human Genome Project: A collection of projects conducted by scientists around the world to map, or identify, all human genes.

Humulin: A commercial insulin, or artificial hormone, made by genetic engineering.

hybrid: The offspring of breeding plants or animals who are of different genetic categories.

immune system: The integrated system of organs, tissues, cells, and cell products that neutralize harmful organisms and substances.

interferons: Any of a group of proteins produced by cells to prevent a virus from replicating and causing damage.

marker: A known DNA sequence associated with a particular gene that is used to indicate the presence of a gene or trait.

microorganisms: An organism of microscopic or submicroscopic size.

molecular biology: The branch of biology that deals with molecules essential to life, such as nucleic acids and proteins, and their role in cell replication and transfer of genetic information.

molecular pharming: The use of genetically engineered animals to produce pharmaceuticals for humans, as if the animals were farms.

mutation: A sudden alteration in the nucleotide sequence of the DNA coding for a gene and thus a physical rearrangement of a chromosome.

National Institutes of Health (NIH): Research institutions, organized and funded by the U.S. Department of Health and Human Services.

nucleotide: The basic structural unit of DNA.

oncogenes: A gene that causes the transformation of normal cells into cancerous tumor cells.

OncoMouse: Harvard University's patented mouse engineered with genes that increase its risk of cancer.

organ: A part of an organism with a special function, such as the eye, liver, heart.

organism: A form of life, such as a plant, an animal, a bacterium, or a fungus.

patent: A grant from the government that allows the creator of an invention the sole right to make, use, and sell that invention for a set period of time.

pesticide: A substance, usually a chemical, that kills pests.

pharmaceuticals: Products manufactured to use as medicinal drugs.

porcine: A growth hormone fed to pigs.

probe: A chemical compound designed to attract chemicals in a portion of one DNA strand and then attach itself to the strand.

recombinant DNA: Genetically engineered DNA prepared by transplanting or splicing genes from one species into the cells of a host organism of a different species.

screen: To examine a person's DNA for the presence of defective genes.

somatic cells: Cells that contain genes controlling an organism's traits, but not future generations.

toxicant, toxin: A poisonous substance.

transgenic: Having genes transferred from another breed or species.

vaccine: A weakened form of a disease that causes the body to develop antibodies but never acquire the disease.

virus: Any of various simple submicroscopic parasites of plants, animals, and bacteria that often cause disease.

Organizations
to Contact

Ag Bioethics Forum
115 Morrill Hall
Iowa State University
Ames, IA 50011
(515) 294-4111

The forum is an interdisciplinary group that focuses on the relationship between agriculture and bioethics. It explores the ethical dilemmas that arise when genetic engineering is applied to agriculture. The forum publishes the newsletter *Ag Bioethics Forum.*

American Civil Liberties Union (ACLU)
132 W. 43rd St.
New York, NY 10036
(212) 944-9800

The ACLU champions the civil rights granted by the U.S. Constitution. It is becoming increasingly concerned with the effects of genetic engineering on the right to privacy and the rights of defendants in criminal trials. The ACLU publishes a variety of handbooks, pamphlets, reports, and newsletters, including the quarterly *Civil Liberties* and the monthly *Civil Liberties Alert.*

American Society of Law, Medicine, and Ethics
765 Commonwealth Ave., Suite 1634
Boston MA 02215
(617) 262-4990
fax: (617) 437-7596

The society's members include physicians, attorneys, health care administrators, and others interested in the relationship between law, medicine, and ethics. It takes no positions but acts as a forum for discussion of issues such as genetic engineering. The organization has an information clearinghouse and a library. It publishes the quarterlies *American Journal of Law and Medicine* and *Journal of Law, Medicine, and Ethics*; the periodic *ASLM Briefings*; and books.

B.C. Biotechnology Alliance (BCBA)

1122 Mainland St., Suite 450
Vancouver, BC V6B 5L1
CANADA
(604) 689-5602
fax:(604) 689-5603
Web site: http://www.biotech.bc.ca/bcba/

BCBA is a nonprofit trade association for producers and users of biotechnology. The alliance works to increase public awareness and understanding of biotechnology, including the awareness of its potential contributions to society. The alliance's publications include the bimonthly newsletter *Biofax* and the annual *Directory of BC Biotechnology Capabilities*.

Biotechnology Industry Organization (BIO)

1625 K St. NW, Suite 1100
Washington, DC 20006
(202) 857-0244
fax: (202) 857-0237

BIO is composed of companies engaged in industrial biotechnology. It monitors government actions that affect biotechnology and through its educational activities and workshops promotes increased public understanding of biotechnology. Its publications include the bimonthly newsletter *BIO Bulletin*, the periodic *BIO News*, and the book *Biotech for All*.

Council for Responsible Genetics
5 Upland Rd., Suite 3
Cambridge, MA 02140
(617) 868-0870
fax: (617) 864-5164

The council is a national organization of scientists, health professionals, trade unionists, women's health activists, and others who want to make sure that biotechnology is developed safely and in the public interest. The council publishes the bimonthly newsletter *GeneWatch* and position papers on the Human Genome Initiative, genetic discrimination, germline modifications, and DNA-based identification systems.

Environmental Protection Agency (EPA)
401 M St. SW
Washington, DC 20460
(202) 260-4700

The EPA administers federal environmental policies, conducts research, enforces regulations, and provides information on many environmental subjects, including biotechnology. Its Pesticides and Toxic Substances division studies how the bacterial by-products and industrial chemicals produced by biotechnology affect the environment. The agency publishes the *EPA Journal* and many other publications on biotechnology and the environment.

Foundation on Economic Trends
1130 17th St. NW, Suite 630
Washington, DC 20036
(202) 466-2823
fax: (202) 429-9602

The foundation examines the environmental, economic, and social consequences of genetic engineering. It believes society should use extreme caution in implementing genetic technologies because it fears that the unwise use of these technologies threatens people, animals, and the environment. The foundation publishes the books *Biological Warfare: Deliberate Release of Microorganisms* and *Reproductive Technology* as well as articles and research reports.

Friends of the Earth
218 D St. SE
Washington, DC 20003
(202) 544-2600
fax: (202) 543-4710

Friends of the Earth monitors legislation and regulations that affect the environment. It speaks out against what it perceives as the negative impact biotechnology can have on farming, food production, genetic resources, and the environment. Friends of the Earth publishes the quarterly newsletter *Atmosphere* and the magazine *Friends of the Earth/Not Man Apart.*

Genetic Society of America
9650 Rockville Pike
Bethesda, MD 20814
(301) 571-1825
fax: (301) 530-7079

The society promotes professional cooperation among people working in genetics and related sciences. It publishes the monthly journal *Genetics.*

Health Resources and Services Administration
Department of Health and Human Services
Genetic Services
5600 Fishers Lane
Rockville, MD 20857
(301) 443-1080
fax: (301) 443-4842

The administration provides funds to help develop or enhance regional and state genetic screening, diagnostic counseling, and follow-up programs. It provides funds to develop community-based psychological and social services for adolescents with genetic disorders. It offers many publications through its educational programs, and it produces directories and bibliographies on human genetics.

Kennedy Institute of Ethics
Georgetown University
1437 37th St. NW
Washington, DC 20057
(202) 687-8099
library: (800) 633-3849
fax: (202) 687-6779

The institute sponsors research on medical ethics, including ethical issues surrounding the use of recombinant DNA and human gene therapy. It supplies the National Library of Medicine with an online database of bioethics and publishes an annual bibliography in addition to reports and articles on specific issues concerning medical ethics.

March of Dimes Birth Defects Foundation
1901 L St. NW, Suite 260
Washington, DC 20036
(202) 659-1800
fax: (202) 296-2964

The March of Dimes is concerned with preventing and treating birth defects, including those caused by genetic abnormalities. It monitors legislation and regulations that affect health care and research; awards grants for research; provides funding for treatment of birth defects; offers information on a wide variety of genetic diseases and their treatments; and publishes the quarterly newsletter *Genetics in Practice.*

National Institutes of Health (NIH)
Health and Human Services Dept.
Human Genome Research
9000 Rockville Pike
Bethesda, MD 20892
(301) 402-0911
fax: (301) 402-0837

The NIH plans, coordinates, and reviews the progress of the Human Genome Project and works to improve techniques for cloning, storing, and handling DNA. It offers a variety of information on the Human Genome Project.

U.S. Department of Agriculture
Grants and Program Systems
901 D St. SW
Washington , DC 20250
(202) 401-1761
fax: (202) 401-6488

This division of the U.S. Department of Agriculture administers grants for biotechnology research and oversees such research. It has numerous publications on agriculture and biotechnology.

Suggestions for Further Reading

William Bains, *Genetic Engineering for Almost Everybody*. New York: Viking Penguin, 1987.

Jerry E. Bishop and Michael Waldholz, *Genome: The Story of the Most Astonishing Scientific Adventure of Our Time—The Attempt to Map All the Genes in the Human Body*. New York: Simon and Schuster, 1990.

Richard P. Brennan, *Levitating Trains and Kamikaze Genes: Technological Literacy for the 1990s*. New York: John Wiley & Sons, 1990.

Joel Davis, *Mapping the Code: The Human Genome Project and the Choices of Modern Science*. New York: John Wiley & Sons, 1990.

Sylvia Engdahl and Rick Robertson, *Tools for Tomorrow: New Knowledge About Genes*. New York: Atheneum, 1979.

D. S. Halacy Jr., *Genetic Revolution: Shaping Life for Tomorrow*. New York: Harper & Row, 1974.

Richard Hutton, *Bio-Revolution: DNA and the Ethics of Man-Made Life*. New York: New American Library, 1978.

Christopher Lampton, *DNA and the Creation of New Life*. New York: Arco Publishing, 1983.

Thomas F. Lee, *The Human Genome Project: Cracking the Genetic Code of Life*. New York: Plenum Press, 1991.

Margery L. Oldfield, *The Value of Conserving Genetic Resources*. Sunderland, MA: Sinauer Associates, 1984.

Steve Olson, *Biotechnology: An Industry Comes of Age.* Washington, DC: National Academic Press, 1986.

Brian Stableford, *Future Man.* New York: Crown Publishers, 1984.

Eva Stwertka and Albert Stwertka, *Genetic Engineering.* New York: Franklin Watts (an Impact Book), 1982.

Edward Sylvester, *Genetic Engineering and the Next Industrial Revolution.* New York: Charles Scribner's Sons, 1989.

Burke K. Zimmerman, *Biofuture: Confronting the Genetic Era.* New York: Plenum Press, 1984.

Works Consulted

William H. Allen, "Farming for Spare Body Parts," *BioScience*, February 1995.

Natalie Angier, "Microbe DNA Seen as Alien by Immune Cells," *The New York Times*, April 11, 1995.

———, "With New Fly, Science Outdoes Hollywood," *The New York Times*, March 23, 1995.

Sandra Blakeslee, "Overproduction of a Protein Is Linked to Adult's Diabetes," *The New York Times*, February 2, 1995.

Mark Caldwell, "The Fix Sticks," *Discover*, January 1995.

———, "Prokaryotes at the Gate," *Discover,* August 1994.

Liebe F. Cavalieri, "Ecological Brinkmanship: The Genetic Engineering of Microbes and Plants," *The World & I*, December 1994.

The Chromosome's Core," *Discover*, May 1994.

Marla Cone, "The Mouse Wars Turn Furious," *Los Angeles Times*, May 9, 1993.

Congress of the United States, Office of Technology Assessment, *Mapping Our Genes: Genome Projects: How Big, How Fast?* Baltimore: The Johns Hopkins University Press, 1988.

Elizabeth Culotta and Daniel E. Koshland Jr., "DNA Repair Works Its Way to the Top," *Science*, December 23, 1994.

Edgar J. DaSilva and Albert Sasson, "Achievements, Expectations, and Challenges," *The UNESCO Courier*, June 1994.

Jean Dausset, "Scientific Knowledge and Human Dignity," *The UNESCO Courier*, September 1994.

Bernard D. Davis, ed., *The Genetic Revolution: Scientific Prospects and Public Perceptions*. Baltimore: The Johns Hopkins University Press, 1991.

Renato Dulbecco, "The Prospects for Gene Therapy," *The UNESCO Courier*, September 1994.

"Facing the Future: Genetic Engineering," *The Animals' Agenda*, January/February 1995.

Hans Galjaard, "Prenatal Diagnosis: Foretelling the Quality of Life," *The UNESCO Courier*, September 1994.

"Gene Implanted in Mice via Pregnant Mother," *The New York Times*, February 28, 1995.

"Gene Is Shown to Block Spread of Prostate Cancer in Laboratory Mice," *The New York Times*, May 12, 1995.

"Gene May Keep Arteries of Heart Patients Open," *The New York Times*, January 27, 1995.

Josie Glausiusz, "Flap over Milk," *Discover*, January 1995.

———, "A Gene for Breast Cancer," *Discover,* January 1995.

David Graham, "Redesigning the Plant," *San Diego Union-Tribune*, September 28, 1994.

Joann Gutin, "End of the Rainbow," *Discover*, November 1994.

Stephen S. Hall, "Protein Images Update Natural History," *Science*, February 3, 1995.

Will Hively, "Married to the Molecule," *Discover*, October 1994.

Robert Lee Hotz, "Fruits of Genetic Tinkering Are Headed for U.S. Tables," *Los Angeles Times*, November 12, 1993.

George H. Kieffer, *Biotechnology, Genetic Engineering, and Society.* Reston, VA: National Association of Biology Teachers, 1987.

Gina Kolata, "Alzheimer's Is Produced in Mice, Report Says," *The New York Times*, February 9, 1995.

———, "Gene Therapy Fails to Yield Any Benefits, 2 Studies Find," *The New York Times*, September 28, 1995.

———, "In the Rush Toward Gene Therapy, Some See a High Risk of Failure," *The New York Times*, July 25, 1995.

———, "Molecular Tools May Offer Clues to Reducing Risks of Birth Defects," *The New York Times*, May 23, 1995.

———, "Mystery Is Lifting on Mechanisms of Disease That Killed Lou Gehrig," *The New York Times*, May 9, 1995.

———, "Tests to Assess Risks for Cancer Raising Questions," *The New York Times*, March 27, 1995.

———, "A Vat of DNA May Become Fast Computer of the Future," *The New York Times*, April 11, 1995.

Georges B. Kutukdjian, "UNESCO and Bioethics," *The UNESCO Courier*, September 1994.

Scott LaFee, "The Future Harvest: Promise, Menace of Biotech Create Sharp Debate," *San Diego Union-Tribune*, September 28, 1992.

Warren E. Leary, "Research Hints of Immunization by Fruits," *The New York Times*, May 5, 1995.

"Leukemia Treatment Found to Work in Mice," *The New York Times*, February 10, 1995.

Thomas H. Maugh II, "Costly New Treatments Put Insurers in Quandary," *Los Angeles Times*, March 28, 1994.

———, "Major Step in Gene Therapy Reported," *Los Angeles Times*, May 10, 1995.

———, "Pace of Gene Therapy Quickens," *Los Angeles Times*, November 2, 1993.

———, "Unraveling the Secrets of Genes," *Los Angeles Times*, October 31, 1993.

Rosie Mestel, "Ascent of the Dog," *Discover*, October 1994.

"A More Useful Mouse," *Discover*, December 1994.

"Noëlle Lenoir Talks to Bahgat Elnadi and Adel Rifaat," *The UNESCO Courier*, September 1994.

G. J. V. Nossal, *Reshaping Life: Key Issues in Genetic Engineering*. New York: Press Syndicate of the University of Cambridge, 1985.

Joe Palca, "The Promise of a Cure," *Discover*, June 1994.

Peter Radetsky, "Speeding Through Evolution," *Discover*, May 1994.

William H. Safer, *Introduction to Genetic Engineering*. Boston: Butterworth-Hernemann, 1991.

Jane K. Setlow, ed., *Genetic Engineering: Principles and Methods*. Vol. 15. New York: Plenum, 1993.

Ruth SoRelle, "The Gene Doctors: Seeking Cures on Medicine's Frontiers," *Houston Chronicle*, April 2, 1995.

Robert Steinbrook, "The Promise and Perils of New Genetic Screening," *Los Angeles Times*, November 1, 1993.

Sheryl Stolberg, "Insurance Falls Prey to Genetic Bias," *Los Angeles Times*, March 27, 1994.

Ezra N. Suleima, "Genetics in the Market-Place," *The UNESCO Courier*, September 1994.

David Suzuki and Peter Knudtson, *Genetics: The Clash Between the New Genetics and Human Values*. Cambridge, MA: Harvard University Press, 1989.

Kathy A. Svitil, "Defending the Joints," *Discover*, October 1994.

Monkombu Sambasivan Swaminathan, "Biotechnology for Beginners," *The UNESCO Courier*, June 1994.

"Switched-On Green," *Discover*, June 1994.

Indra K. Vasil and Luis Herrera-Estrel, "From the Green Revolution to the Gene Revolution," *The UNESCO Courier*, June 1994.

Nicholas Wade, "Rapid Gains Are Reported on Genome," *The New York Times*, September 28, 1995.

Juan Williams, "Violence, Genes, and Prejudice," *Discover*, November 1994.

Index

About the Author

After many years of teaching British literature, Clarice Swisher now devotes her time to research and writing. She is the author of *The Beginning of Language, Relativity, Albert Einstein, Pablo Picasso, The Ancient Near East,* and *The Glorious Revolution.* She is currently working on a new series of literary companions to American, British, and world authors. She lives in Saint Paul, Minnesota.

Picture Credits